Engineering Projects for Kids
2 Books in 1

Soccer Inspired Physics and Engineering Activities for Kids

Table of Contents

Table of Contents 3
Free Gift 5
DISCLAIMER 6
Introduction 7
Conservation of Momentum in Soccer 9
The Beauty of a Ground Pass 21
Flight of the Soccer Balls: The Projectile Trajectory 32
How The Soccer Wind Can Blow Your Mind 42
How a Soccer Ball Bends Through the Air 49
How the Goalkeeper Shuts You Down – The Goalkeeping Glove Impact 57
Angled Shots: The More You See, the More you Score. 64
Bonus: What a Drag 85
Conclusion 88
Challenge Answers 90
Book 2 108
Free Gift 109
Introduction 110
The Search for that Elusive Banana 112
Taxi Please 129
Number Key Challenge 149
Loops 163
Nested Loops 173

Conditional Loops 181

Programming Challenges 186

Answers... 195

The Search for that Elusive Banana 196

Taxi Please... 210

Number Key Challenge - Answers... 220

Loops... 223

Nested Loops... 228

Conditional Loops 232

Programming Challenges 235

Free Gift

We do want you to succeed in coding. To ensure your success, we are giving you a free list of projects that you can work on once you are completed with this book.

https://coding.gr8.com/

DISCLAIMER

Copyright © 2023

All Rights Reserved

No part of this eBook can be transmitted or reproduced in any form, including print, electronic, photocopying, scanning, mechanical, or recording, without prior written permission from the author.

While the author has taken the utmost effort to ensure the accuracy of the written content, all readers are advised to follow the information mentioned herein at their own risk. The author cannot be held responsible for any personal or commercial damage caused by the information. All readers are encouraged to seek professional advice when needed.

Introduction

Soccer, or football as it is recognised in many parts of the world, is the most popular sport on the planet. Millions of fans around the globe watch great players and teams compete at the highest level, with the FIFA World Cup being one of the most broadly watched events in the world. Beyond the excitement of the game, there are endless lessons to be learnt from soccer that can be utilized to different areas of life, including engineering.

Engineering is a vast and complex field that encompasses a wide variety of disciplines, from mechanical and electrical engineering to materials science. At its core, engineering is about problem solving that make our lives better. Whether it's designing a new product, building a bridge, or creating a software program application, engineering requires creativity, critical thinking, and a deep appreciation for the standards and techniques that underpin the field. Both soccer and engineering require a great deal of teamwork, adaptability and goal setting, in addition to skill, motivation and dedication.

In this book, we'll focus on a few basic principles of engineering that apply to soccer. We investigate how engineering principles affect the behavior of the soccer ball, soccer gloves and the techniques of soccer players. This book is great for soccer fans to learn more about their favorite sport, and also for those who want to better understand some of the basic principles in engineering.

Conservation of Momentum in Soccer

If you watch or play soccer, you might wonder why a ball moves the way it does. Why does the ball move faster in some instances and go farther? Why does it bounce higher sometimes? In this chapter, we'll take a closer look at the behavior of the soccer ball after it is kicked. We'll use physics to determine the velocity of the ball when it is kicked. And also have a look at other factors that can affect the trajectory of the ball.

First, let's look at momentum.

The momentum of an object is its mass times velocity.

$$p = mv$$

Where p is the momentum, m is the mass and v is the velocity.

When a player kicks a soccer ball, he transfers momentum onto the soccer ball. Like everything in nature, soccer also follows the laws of physics, and in this case, the law of conservation of momentum.

According to the law of conservation of momentum, in a closed system, when two objects collide, the total momentum remains the same.

So, when a player kicks a ball, the total momentum of the player and ball before the kick is equal to the total momentum of the player and ball after the kick.

$$p_{i,ball} + p_{i,player} = p_{f,ball} + p_{fi,player}$$

$p_{i,ball}$ = Initial momentum of ball

$p_{i,player}$ = Initial momentum of player

$p_{f,ball}$ = Final momentum of ball

$p_{f,player}$ = Final momentum of player

Before Kick

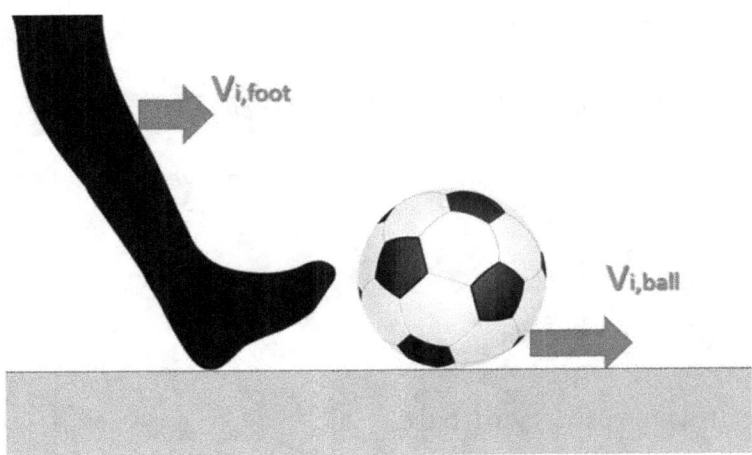

After Kick

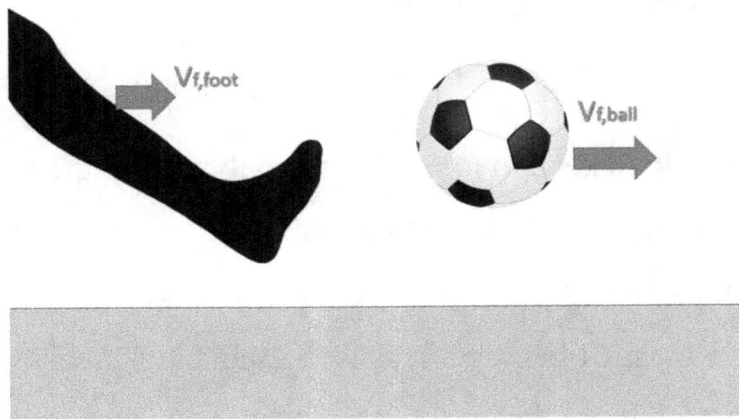

Using the first momentum equation and plugging into the 2nd equation, we get:

$$m_{ball} * v_{i,ball} + m_{player} * v_{i,player} = m_{ball} * v_{f,ball} + m_{player} * v_{f,player}$$

m_{ball} = Mass of ball

m_{player} = Mass of player

$v_{i,ball}$ = Initial velocity of ball

$v_{i,player}$ = Initial velocity of player

$v_{f,ball}$ = Final velocity of ball

$v_{f,player}$ = Final velocity of player

But let's keep in mind that the only part of the player's body kicking the ball is the foot. So, we care about the mass and velocity of the foot. We also use a term known as effective mass of the foot. This is the standalone mass of the portion of the foot that strikes the soccer ball. There are many forces generated by the muscles during the kick that gives it a higher mass than the other standing foot. So, the equation now becomes:

$$m_{ball} * v_{i,ball} + m_{foot,effective} * v_{i,foot} = m_{ball} * v_{f,ball} + m_{foot,effective} * v_{f,foot}$$

m_{ball} = Mass of ball

$m_{foot,effective}$ = Effective mass of foot

$v_{i,ball}$ = Initial velocity of ball

$v_{i,foot}$ = Initial velocity of foot

$v_{f,ball}$ = Final velocity of ball

$v_{f,foot}$ = Final velocity of foot

Now, let's look at a simple example.

The mass of a soccer ball is 410 grams (0.41 kg). It is at rest, so it has an initial velocity of zero.

Let's assume that a soccer player kicks a ball. Let's assume his foot has an effective mass of 1 kg and his foot is moving at a velocity of 20 m/s.

The velocity of the foot after kicking the ball is 5 m/s. So, let's use this data in above formula to calculate velocity of the ball.

Before Kick

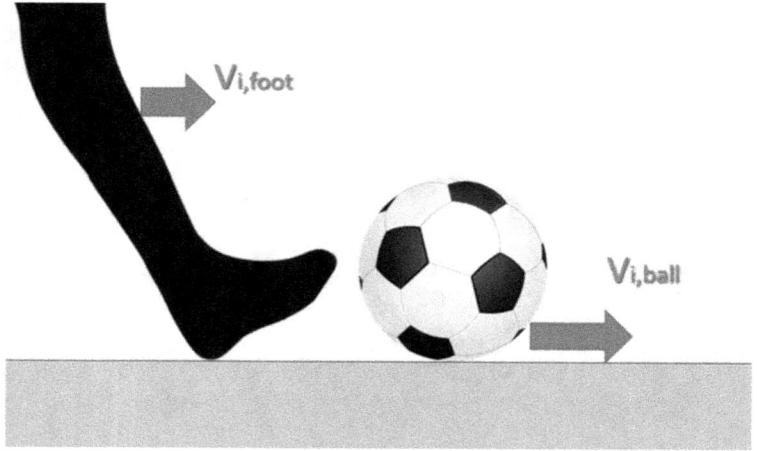

After Kick

$$m_{ball} * v_{i,ball} + m_{foot,effective} * v_{i,foot}$$
$$= m_{ball} * v_{f,ball} + m_{foot,effective} * v_{f,foot}$$

$m_{ball} = 0.41$ kg

$m_{foot,effective} = 1$ kg

$v_{i,ball} = 0$ m/s

$v_{i,foot} = 20$ m/s

$v_{f,ball} = ??$

$v_{f,foot} = 5$ m/s

$$0.41 * 0 + 1 * 20 = 0.41 * v_{f,ball} + 1 * 5$$

$$v_{f,ball} = 36.6 \; m/s$$

As we can see, the final velocity of the ball is 36.6 m/s as some of the momentum from the foot is transferred to the ball.

Elasticity

Now, let's look at the conservation of momentum equation again… "According to the law of conservation of momentum, in a closed system, when two objects collide, the total momentum remains the same."

When we did the previous calculations, we assumed that the player and ball are in a "closed system". But is that true? A closed system means nothing else is interacting between the player and ball. But in real life, there is friction, heat loss, wind etc. that affects the momentum of the ball. Friction is two surfaces rubbing against each other.

What happens is that there is a slight loss in momentum due to above factors. So, the final momentum is slightly less than the initial momentum. This loss in momentum is indicated by a factor known as elasticity. In equation below, the initial momentum is multiplied by an elasticity factor e that is less than 1.

$$(p_{i,ball} + p_{i,player}) * e = (p_{f,ball} + p_{fi,player})$$

If e=1; it indicates a perfectly inelastic collision between player and ball. However, in real life e is less than 1.

$$(m_{ball} * v_{i,ball} + m_{foot,effective} * v_{i,foot}) * e = e * (m_{ball} * v_{f,ball} + m_{foot,effective} * v_{f,foot})$$

Let's look at same example above taking into account an elasticity e of 0.95.

Before Kick

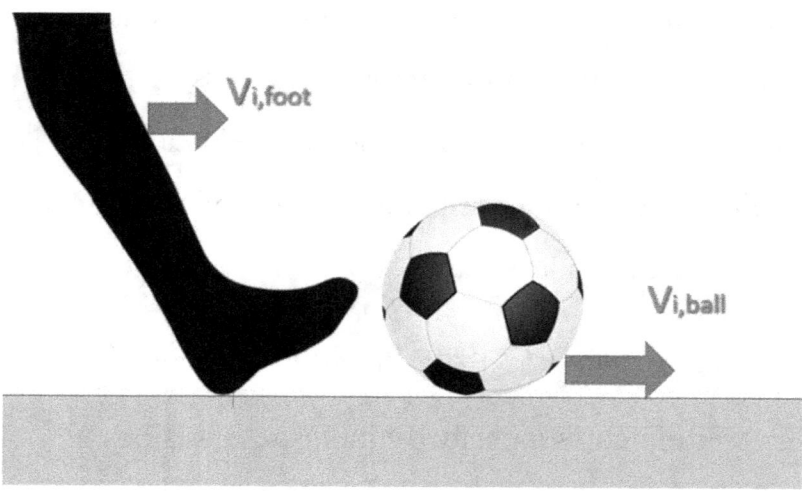

After Kick

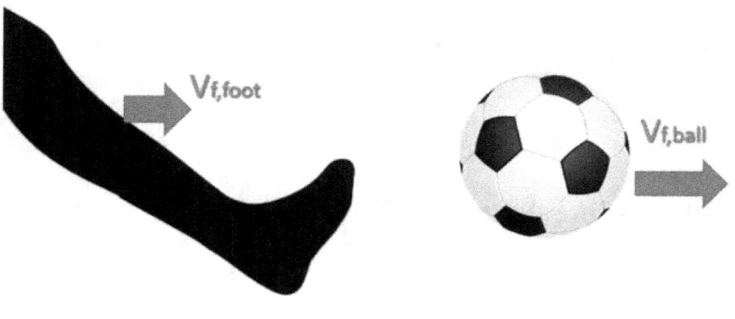

$m_{ball} = 0.41$ kg

$m_{foot, effective} = 1$ kg

$v_{i,ball} = 0$ m/s

$v_{i,foot} = 20$ m/s

$v_{f,ball} = ??$

$v_{f,foot} = 5$ m/s

$$(0.41 * 0 + 1 * 20) * 0.95 = (0.41 * v_{f,ball} + 1 * 5)$$

$$v_{f,ball} = 34.14 \, m/s$$

The final velocity of the ball is **34.14 m/s** which is lower than the first calculation of 36.6 m/s. This is because some of the energy is lost to the environment due to heat, friction etc. This is because the player-ball system is not a completely open system, and we need to take elasticity into account.

Now, here's an example of you to practice:

Challenge 1:

The mass of a soccer ball is 410 grams (0.41 kg). It is moving slowly, with an initial velocity of 2 m/s in the same direction of the kick.

The soccer player kicks a ball. His foot has an effective mass of 1 kg, and his foot is moving at a velocity of 10 m/s.

The velocity of the foot after kicking is 7 m/s. The elasticity of the system is 0.92. Calculate the final velocity of the ball.

The Beauty of a Ground Pass

A ground pass is among the most fundamental skills in soccer. A team that plays a ground passing game is the best to watch. An Arsenal, Barcelona or Manchester City seems more attractive to watch than a Liverpool. The beauty lies in the effectiveness, simplicity and creativity used to unlock the tightest of defences and create goal scoring opportunities. It requires a precise weighting of the pass, close control and fast runs from teammates.

In this section, we'll have a look at how fast a ball moves for a ground pass. We'll look at the forces acting on the ball and how it affects the speed and acceleration of the soccer ball.

Before we get started, let's define Newton's second law of motion, which states that "the acceleration of an object is dependent on two variables, the external force acting on the object and the mass of the object."

In the case of a ground pass, there is a kicking force acting in the horizontal direction (we'll call this the x-axis). And there is no external force acting on the ball in the vertical direction (the y-axis).

To describe all the forces acting on the ball, we use something known as the free body diagram. A free body diagram is a simplified version of all forces acting on an object at a particular moment in time.

The Free body diagram of the soccer ball is below.

In the vertical direction, the weight of the soccer ball W acts downwards, and there is a reaction force F_n that the ground exerts on the ball in the upward direction.

The player exerts a kicking force F_k on the ball to the right in the horizontal direction. This induces a frictional force f_s between the ground and the ball in the opposite direction. Friction is the force generated by two surfaces rubbing against each other.

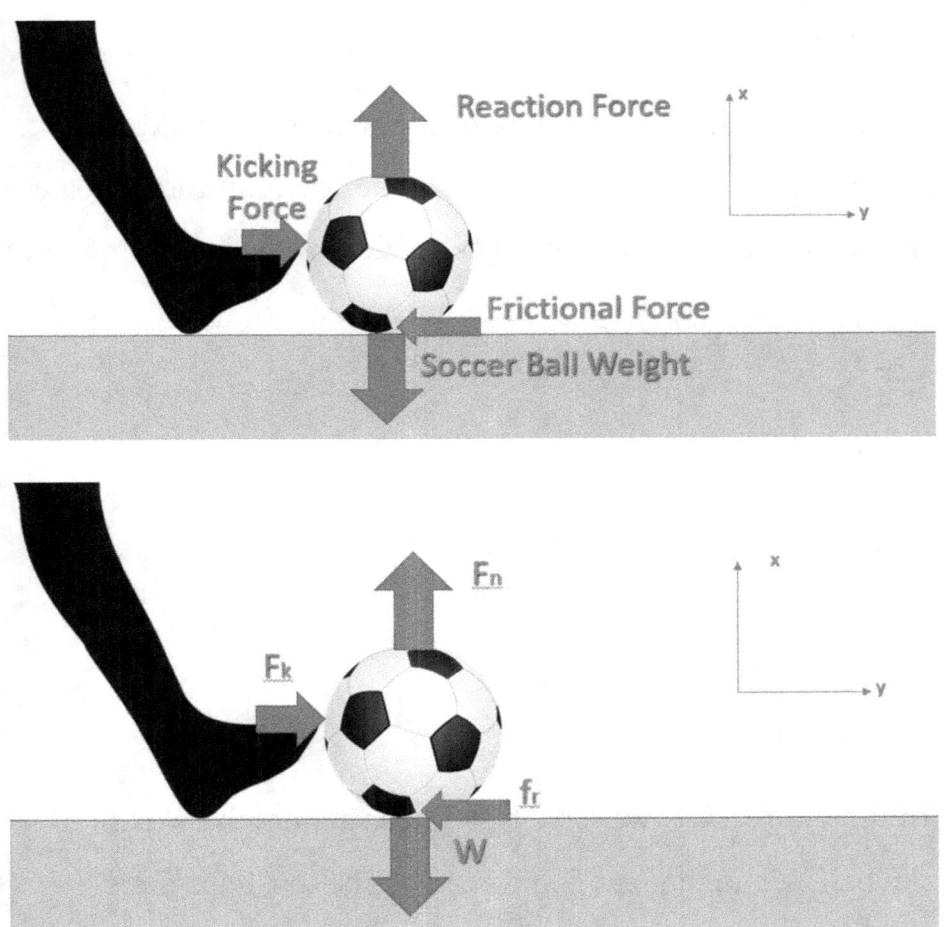

In the free body diagram, we sum all the forces in the two directions to solve the problem.

x-direction:

In the x-direction, we have two forces acting, the kicking force F_k and the frictional force f_s. The resultant force creates motion to the right. Let the acceleration of the ball to the right be a_x. Now, based on Newton's second law of motion, we get:

$$F_k - f_s = m\, a_x$$

Where **m** is the mass of the ball

y-direction:

In the y-direction, we have two forces acting, the weight of the ball W and the reaction force F_n. The resultant force doesn't create any motion up or down, so acceleration in the y-direction is 0. Now, based on Newton's second law of motion, we get:

$$F_n - W = 0$$

$$F_n = W$$

The reaction force is equal and opposite to the weight of the ball.

Now that we know all the forces acting on the ball, let's have a look at each of them individually:

Weight of Ball W

The weight of the ball is the force exerted due to gravity on the ball. So, the force exerted downwards on the ball is the mass of the ball m times the acceleration due to gravity g.

W=m*g

The value of g was determined to be 9.81 m/s² on earth by great philosopher Isaac Newton.

Normal Reaction Force F_n

Now, if there was no force acting upwards on the ball, it would just rip through the earth and fly off into the solar system on the other side. However, there is a force acting upwards that is just enough to keep it stable vertically. The force is not too high to prevent it from moving upwards. This force is known as the normal reaction force F_n.

Frictional Force f_s

When a ball is kicked horizontally on the ground, we know that it eventually comes to rest no matter how hard it is kicked. This is because there is a force acting on the ball in a direction opposite to motion. This force is caused by the interaction between the ball and the grass and is called frictional force. The rougher the surface, the higher the frictional force and the faster the ball stops. That's why you see that the soccer ball stops faster on a sandy surface versus a smooth glass surface.

That's because the sandy surface has a higher coefficient of friction μ. The coefficient of friction μ is a property of the surface. A polished smooth glass surface has a coefficient of friction of almost zero, so you see that the ball barely slows down on the glass surface when it is rolled on a glass surface. The formula for frictional force is below:

$f_s = \mu\, F_n$

The frictional force depends on the coefficient of friction of the surface and the normal reaction of the surface. When we plug in the value of the normal reaction force below, we get the value of f_s:

$f_s = \mu\, W = \mu\, mg$

So, the heavier the ball and the rougher the surface, the higher the frictional force and the faster the ball stops.

Now, let's apply the information to the forces in y-direction:

$F_n = W$

$F_n = m*g$

And, in the x-direction we get:

$F_k - f_s = m\, a_x$

$F_k - \mu\, mg = m\, a_x$

$\mathbf{F_k = \mu\, mg + m\, a_x}$

Or

$a_x = F_k/m - \mu\, g$

So, we see that a larger force creates a larger acceleration along the ground. A higher coefficient of friction and a heavier ball both reduce acceleration.

Now, let's look at a simple example.

Ronaldo kicks a stationary ball along the ground with a force of 150 N. The ball weighs 0.41 kg and they are playing on a soccer field with a coefficient of friction of 0.62 between the grass and the ball. What's the acceleration of the ball?

F_k = Force on the ball = 150 N

m = mass of the ball = 0.41 kg

μ = Coefficient of static friction = 0.62

g = acceleration due to gravity = 9.81 m/s²

Now, using the acceleration equation, we get:

$a_x = F_k/m - \mu g$

   =150/0.41 − 0.61*9.8

   =359.9 m/s²

Now, that we have the acceleration of the kick, what is the velocity of the ball after 2 seconds on the ground?

We know that the acceleration is the change in velocity over time.

$a_x = (v_x - u_x)/t$

where:

v_x = Final velocity after 2 seconds = ??

u_x = Initial velocity of stationary ball = 0

t = time = 2 seconds

Now, let's plug these values into the formula

$a_x = (v_x - u_x)/t$

$359.9 = (v_x - 0)/2$

$v_x = **719.8 \text{ m/s}**$

The velocity after 2 seconds is **719.8 m/s**.

And here's the free body diagram for the kick…

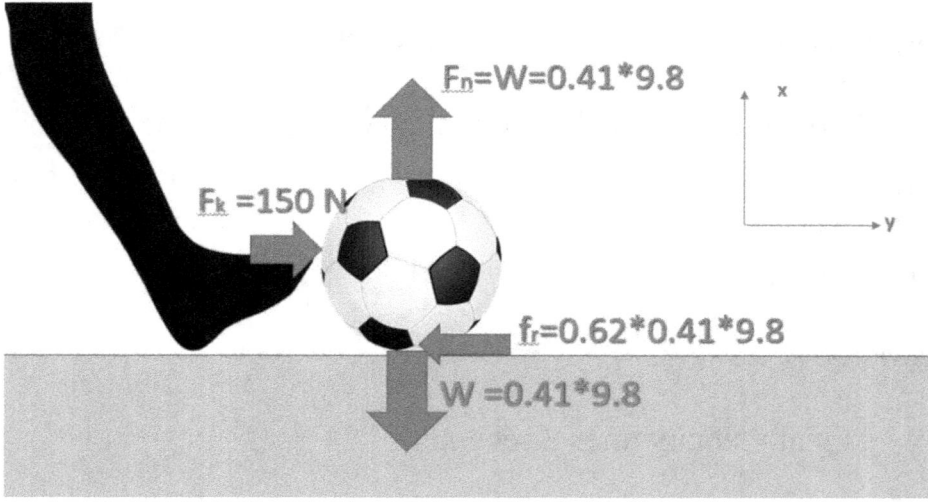

Exercises

Challenge 2

Now, it's your turn.

Messi passes a stationary ball along the ground with a force of 100 N. The ball weighs 0.41 kg and they are playing on a soccer field with a coefficient of friction of 0.7 between the grass and the ball. What's the acceleration of the ball? Fill in the free body diagram below while solving the problem.

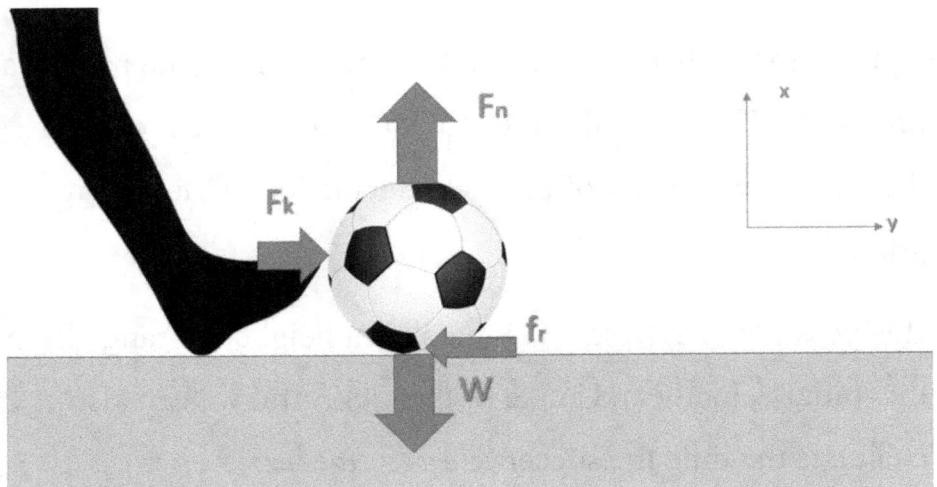

Flight of the Soccer Balls: The Projectile Trajectory

The flight of a soccer ball is an incredibly beautiful sight. The skills of the midfielder who can kick a ball at just the right angle of elevation and pace to catch the run of the offensive player is a marvel to watch. David Beckham was a master of the long pass. There's even a goal he's made from the halfway line.

When a soccer ball is kicked in the air, it follows the path of a projectile. The larger the angle of elevation of the kick, and the faster the velocity of the kick, the higher and farther it goes.

It is possible to calculate the maximum height and range from the initial velocity and angle of the kick. And we can also calculate the time the soccer ball is in the air.

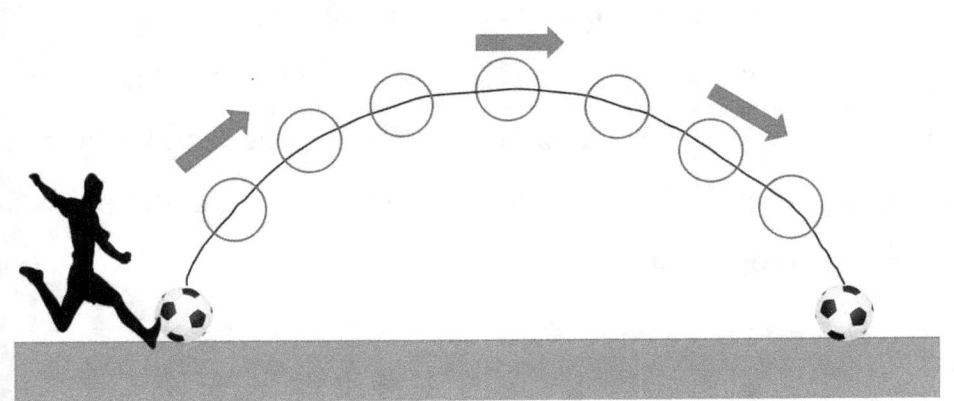

Let's say that the soccer ball is kicked at a speed of u m/s at an angle of elevation of Θ. It follows the projectile trajectory below with a maximum height H and a range of R.

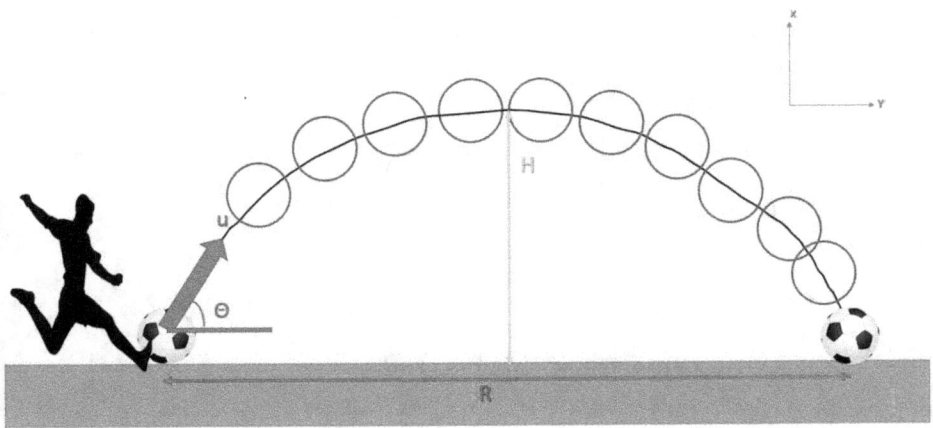

The velocity u can be broken down into horizonal and vertical components using trigonometry. Breaking this down helps us to get a horizontal and vertical component. Let's say that the horizontal component of velocity u is u_x and the vertical component is u_y

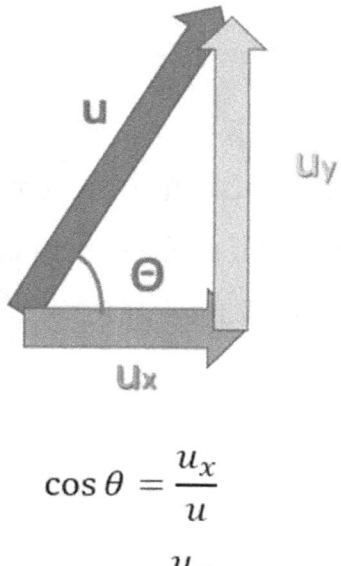

$$\cos \theta = \frac{u_x}{u}$$

$$\sin \theta = \frac{u_y}{u}$$

Now, we get values for u_x and u_y below by cross multiplying:

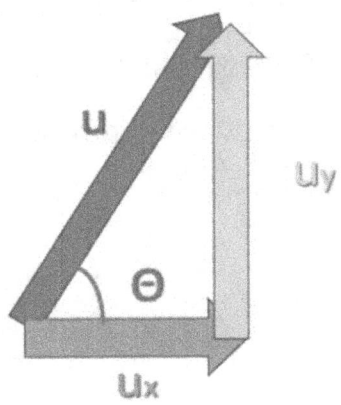

$u_x = u \cos \theta$
$u_y = u \sin \theta$

Now, after inserting these values into the diagram, we get the updated diagram below:

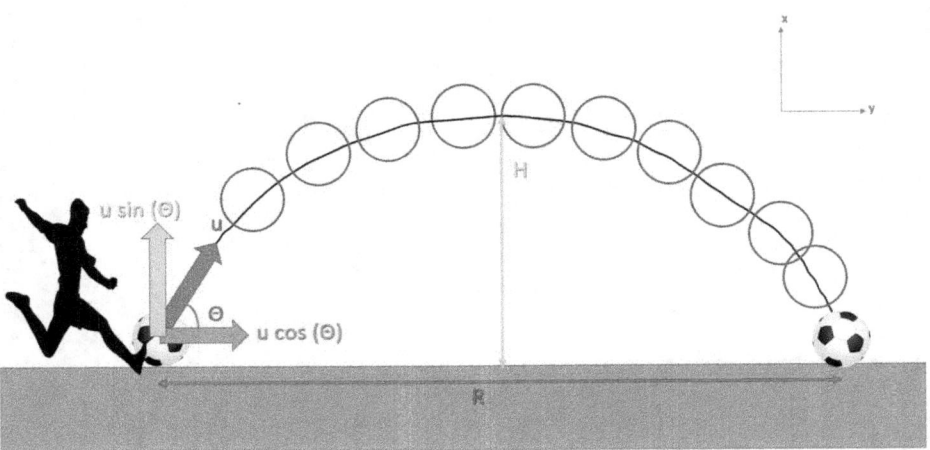

So, once we apply our force calculation in the x-direction and y-direction, we get the below values for horizontal range R and maximum height H. The details of how we get these 3 equations are beyond the scope of this book.

$$H = \frac{u^2(\sin\theta)^2}{2g}$$

$$R = \frac{u^2 \sin 2\theta}{g}$$

$$t = \frac{2u \sin\theta}{g}$$

Now, let's look at an example:

Fred kicks the soccer ball at 20 m/s at an angle of elevation of 30 degrees. How long is the ball in the air, and how long does it travel? What's the maximum height achieved?

To solve this, first let's draw the diagram of the ball with the relevant data.

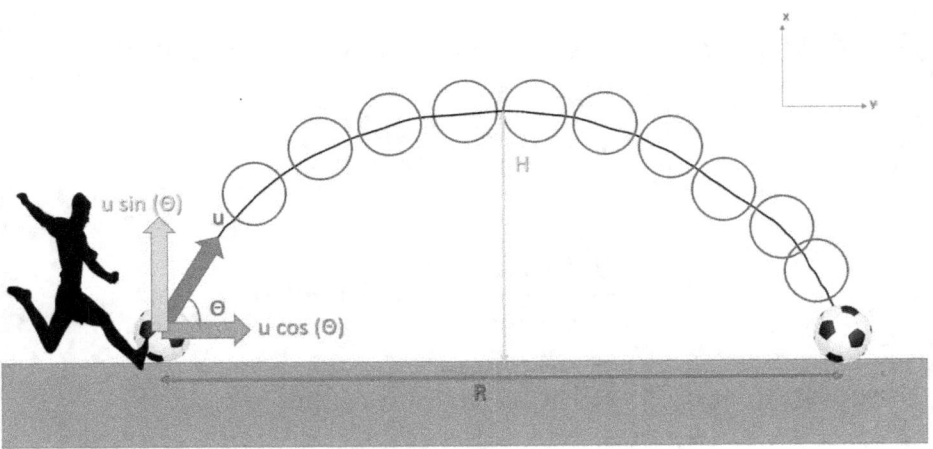

We know from the data that:

u=20 m/s

Θ = 30 degrees

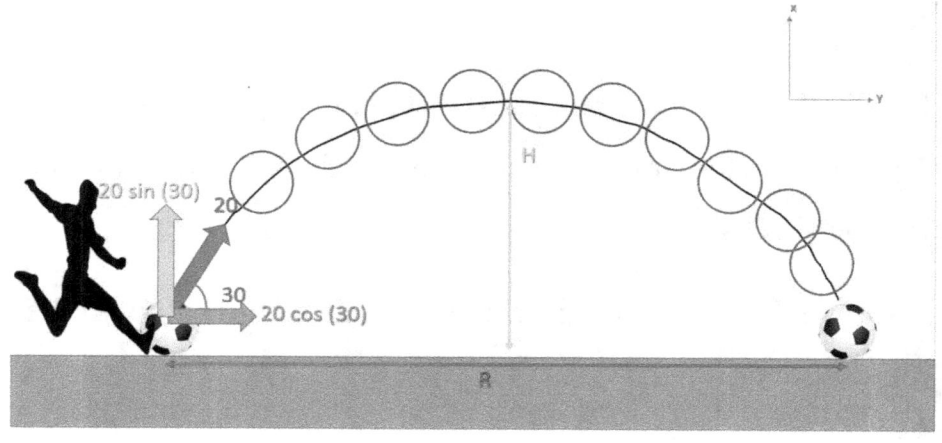

Based on the equation for maximum vertical height, we get:

$$H = \frac{u^2 (\sin\theta)^2}{2g}$$

$$H = \frac{20^2 (\sin 30)^2}{2 * 9.8} = 5.1 \; m \; high$$

Based on the equation for horizontal range, we get:

$$R = \frac{u^2 \sin 2\theta}{g}$$

$$R = \frac{20^2 \sin 60}{9.8} = 35.35 \; m$$

Based on the equation for time of flight we get:

$$t = \frac{2u \sin \theta}{g}$$

$$t = \frac{2 * 20 \sin 30}{9.8} = 2.04 \; seconds$$

So, the ball travels in the air 35.35 m long in 2.04 seconds. It achieves a maximum height of 5.1 m before coming down.

Now, it's time for a practice exercise for you…

Challenge 3

Tim kicks a soccer ball with an initial speed of 10 m/s at an angle of elevation of 45 degrees. Fill out the diagram below and figure out how far the ball goes, and the maximum height it can achieve.

Bonus: Is it going to fly over a goalpost 10 m away? (Hint: The height of a goalpost is 2.7 m)

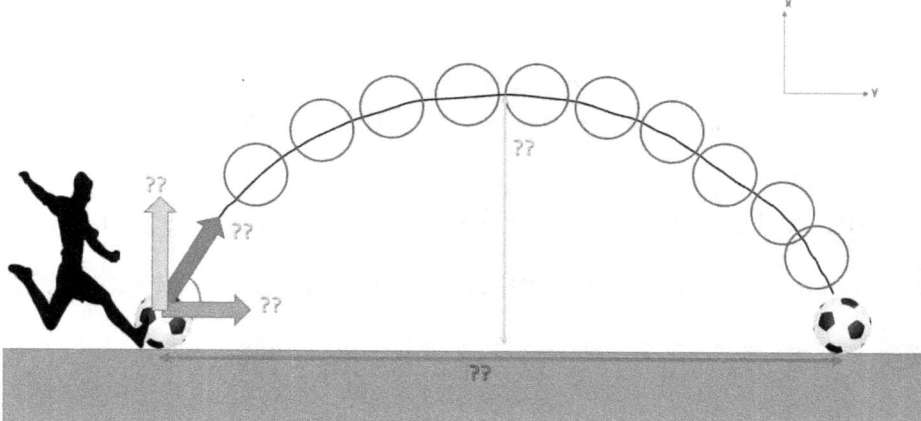

A Short message from the Author:

Hey, are you enjoying the book? I'd love to hear your thoughts!

Many readers do not know how hard reviews are to come by, and how much they help an author.

I would be incredibly thankful if you could take just 60 seconds to write a brief review on Amazon, even if it's just a few sentences!

Browse to the product page and leave a review as shown below.

Thank you for taking the time to share your thoughts!

Your review will genuinely make a difference for me and help gain exposure for my work.

How The Soccer Wind Can Blow Your Mind

In the previous chapter, "Flight of the Soccer Ball" we calculated the projectile motion of a soccer ball kicked with a certain velocity. However, there's one thing missing. The wind.

The wind complicates things a little. If the wind blows in the direction of the soccer ball, it increases the speed of the ball. If the wind blows in the opposite direction of the ball, it decreases the speed of the ball.

There are two terms that we need to understand to move forward, relative velocity and absolute velocity.

Absolute velocity is the velocity of the ball assuming that the wind speed is zero. In the previous chapter, we used only absolute velocity as we didn't consider the wind speed.

Relative velocity is the velocity of the ball considering the velocity of the wind. The wind speed slows down or speeds up the relative velocity of the ball depending on the actual direction of the wind.

Ball moving in direction of wind.

As mentioned, when the ball moves in direction of wind, there is an increase in speed due to the wind.

The relative velocity of the ball is just the velocity of the ball plus the velocity of the wind. This increase in velocity means that the ball travels further horizontally than it otherwise would.

As seen in the diagram below, the relative velocity of the ball ($V_{ball,rel}$) is velocity of the ball (V_{ball}) plus velocity of the wind (V_{wind}).

 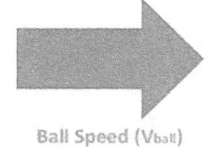 

Ball Speed (V$_{ball}$)

Wind Speed V$_{wind}$

Relative Velocity V$_{ball,rel}$ = V$_{ball}$ + V$_{wind}$
Absolute Velocity V$_{ball,abs}$ = V$_{ball}$

Ball Moving Against the Wind

When the ball is going against the wind, the ball slows down. The relative velocity of the ball decreases by the velocity of the wind. The relative velocity of the wind is just the velocity of the ball V$_{ball}$ minus the velocity of the wind V$_{wind}$.

 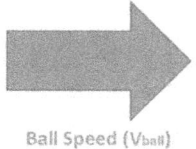

Ball Speed (V$_{ball}$)

Relative Velocity V$_{ball,rel}$ = V$_{ball}$ − V$_{wind}$
Absolute Velocity V$_{ball,abs}$ = V$_{ball}$

This might seem like a simple concept to some, but it's very important to get the simple concepts right before moving on to advanced concepts.

Now, let's look at a few simple examples.

Ball in Direction of Wind - Example

A ball is moving at a velocity of 27 mph in the direction of wind. The wind is blowing at 10 mph. What is the absolute velocity and relative velocity of the ball?

$V_{ball} = 27$ mph

$V_{wind} = 10$ mph

$V_{ball,rel} = ??$

$V_{ball,abs} = ??$

 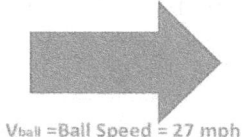 

V_{wind} = Wind Speed = 10 mph

V_{ball} = Ball Speed = 27 mph

$V_{ball,rel}$ = 27+10 = 37 mph
$V_{ball,abs}$ = 27 mph

Ball in Opposite Direction of Wind - Example

A ball is moving at a velocity of 27 mph in the direction of wind. The wind is blowing in the opposite direction at 10 mph. What is the absolute velocity and relative velocity of the ball?

V_{ball} = 27 mph

V_{wind} = 10 mph

$V_{ball,rel}$ = ??

$V_{ball,abs}$ = ??

 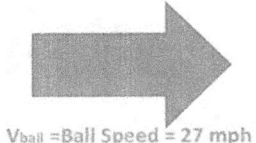 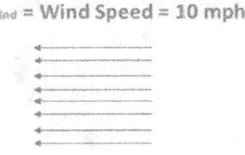

$V_{ball,rel} = 27-10 = 17$ mph
$V_{ball,abs} = 27$ mph

Practical Exercises

Now, it's time for a few practical exercises for you. For each of the below exercises, solve the problem by filling in the diagrams below:

Challenge 4:

The ball is moving at a speed of 20 mph. The wind is blowing against the ball at a speed of 10 mph. What's the absolute and relative velocity of the ball?

 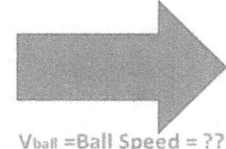 

V_{wind} = Wind Speed = ??

V_{ball} = Ball Speed = ??

Draw Wind above

$V_{ball,rel}$ = ??
$V_{ball,abs}$ = ??

Challenge 5:

A ball is moving at a velocity of 37 mph in the direction of wind. The wind is blowing at 10 mph. What is the absolute velocity and relative velocity of the ball?

V_{wind} = Wind Speed = ??

V_{ball} = Ball Speed = ??

Draw Wind above

$V_{ball,rel}$ = ??
$V_{ball,abs}$ = ??

How a Soccer Ball Bends Through the Air

There's no more beautiful sight in soccer than to see a soccer ball curve in the air towards the top corner of the goal. The goalkeeper thinks the ball is going out due to the trajectory observed. However, it starts curving in midway through the shot and eventually reaches the top corner.

The goalkeeper and defence lie stunned, marvelling at what just happened, while the striker celebrates with his teammates. The skill and precision of the striker is celebrated across the world as a soccer genius. The hours of practice have started to pay off.

The beauty of the moment is due to the interaction between the ball and the atmosphere. The arc of the ball seems to follow the motion of a graceful dance, as it weaves its way across goalkeepers and defenders and finds its way to the top corner. It seems to defy the laws of physics and the laws of gravity.

The Soccer Ball follows a principle known as Bernoulli's theorem, which states that pressure in a fluid (such as air) is inversely proportional to the velocity of the fluid. So, a higher pressure at one point of the fluid means that it has a lower velocity at that point. The difference in pressure causes the ball to move towards the lower pressure surface. It's the same principle on which aeroplanes fly.

In soccer, when a ball is kicked in the air, there is a velocity differential between one side and the other. The side that is moving into the direction of the wind has a lower velocity than the opposite side. This is because the wind must curve more to reach the opposite side. The air facing the wind just glides across the surface at a lower velocity.

According to Bernoulli's theorem, a side with lower velocity has higher pressure and a side with higher velocity has lower pressure. The pressure difference causes a bending force from a direction of higher pressure towards the direction of lower pressure. This force is also called a magnus force and causes the ball to bend. This phenomenon is explained in the diagram below.

See diagrams below.

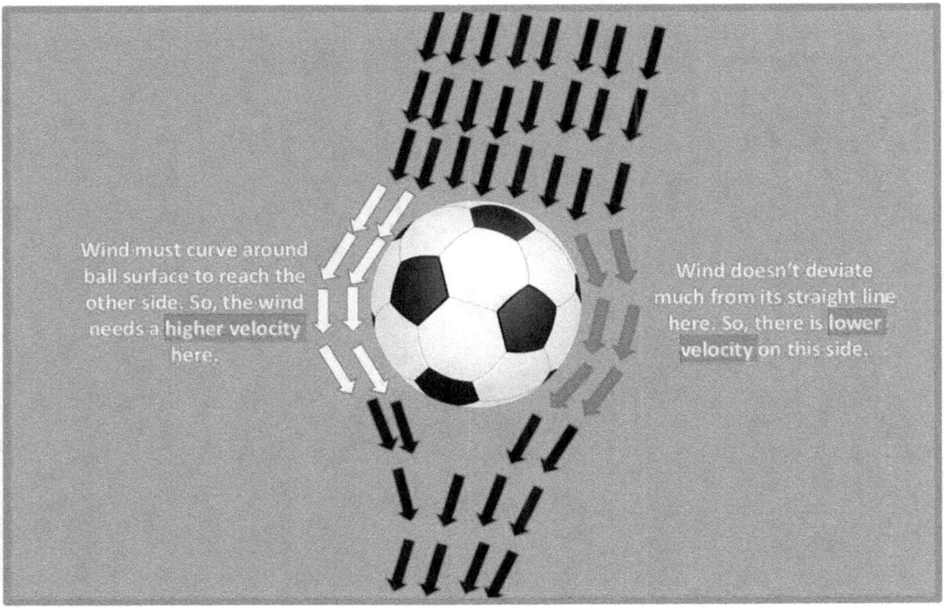

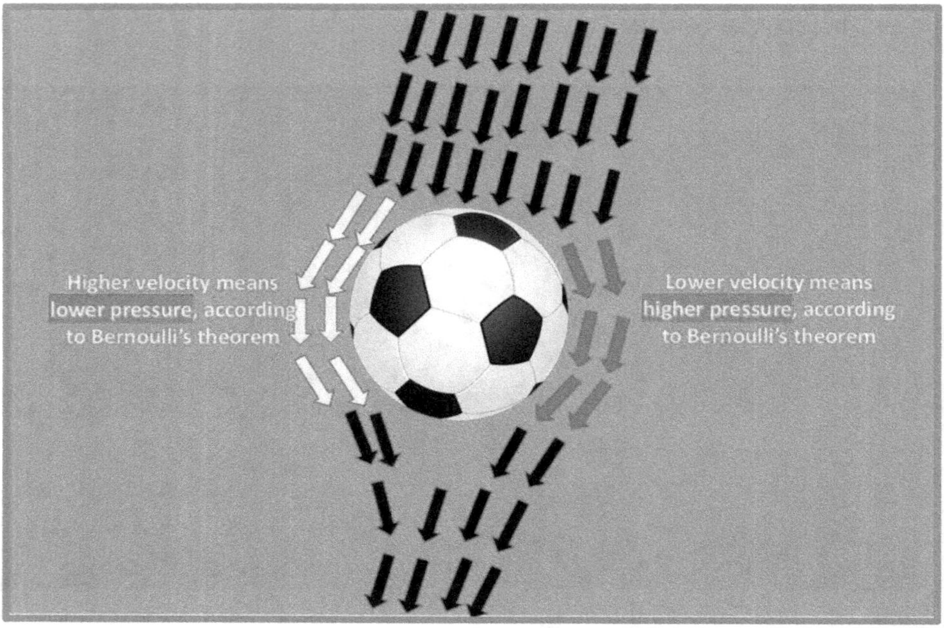

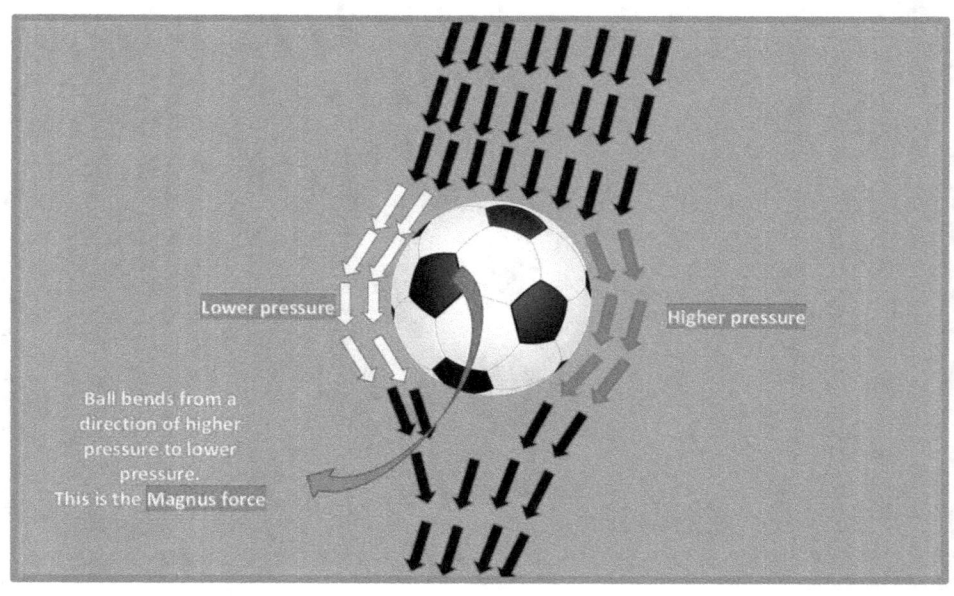

Other factors to consider:

1. **Location of ball contact with foot:** When the ball is just leaving the ground, the area where the ball hits the foot has a higher pressure than the other side. This causes the ball to bend towards the opposite side initially. This is also the dominant factor when there is not much wind around.
2. **Speed of ball:** The faster the ball is moving through the air, the more the Magnus force, and the more it is likely to bend.
3. **Wind speed:** The faster the wind is pushing against the ball, the larger the Magnus force, and the more it is likely to bend.

Exercises:

Challenge 6:

Here's a few fun exercises for you to practice. Here's the ball with the wind flowing against it. In the empty sheet below, draw the areas of high and low pressure and the direction it is likely to bend.

The black areas represent the direction of the wind, and the white arrow is the direction the ball is moving.

Draw your answer in the sheet below:

Challenge 7:

Draw your answer below:

How the Goalkeeper Shuts You Down – The Goalkeeping Glove Impact

The goalkeeper is the reason why you can never seem to score. He's always there, blocking your shot, clearing the ball, punching it, and holding on to it. The goalkeeper is aware of everything that's happening on the soccer field, including movements of the opposing players, and making quick decisions on how to handle opposing players. He can communicate effectively with other players and maintain mental focus all throughout the game.

In this chapter, we're going to have a look at goalkeeping glove design. We see that the top of the glove above the fingers has extra padding. This extra padding makes it easier to punch the ball away. The padding absorbs the impact of the ball moving towards the fingers and offers extra protection to the fingers. Why does this happen? Let's investigate.

As we see below, the ball moves towards the gloves with a velocity and applies a force on the gloves.

From the force equation, we know:

$$F = ma$$

$$F = \frac{m(v - u)}{t}$$

Where F is the force applied on the gloves

m is the mass of the ball

t is the time of contact between ball and gloves

v is final velocity

u is the initial velocity

In the above equation, the soft padding in the gloves increases the time of contact between ball and gloves. This extra time reduces the force that the ball applies on the gloves, and the force that the fingers feel. This increases the protection that the keeper's fingers get.

Now, let's look at an example.

Goalkeeper A decides he has superhuman hands and decides not to wear any gloves. Goalkeeper B is wearing regular goalkeeping gloves with padding. A ball of mass 0.41 kg hits both hands at 20 m/s and both keepers punch the ball back at 10 m/s (the ball is 10 m/s after the punch). The ball contacts Goalkeeper A's hands for 0.1 seconds while it is in contact with Goalkeeper B's gloves for 0.6 seconds. What is the force applied on Goalkeeper A's hands and Goalkeeper B's gloves?

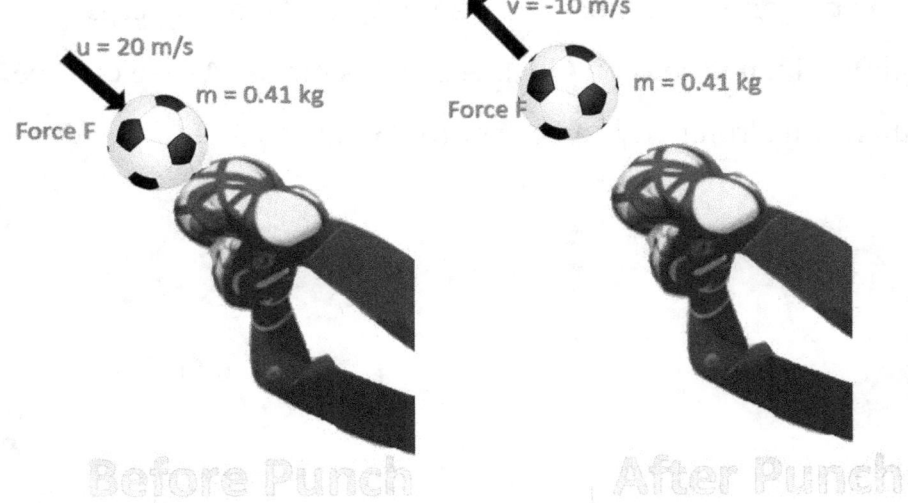

So, we have:

$m = 0.41$ kg

$u = 20$ m/s

$v = -10$ m/s (it goes in opposite direction so -ve sign)

$$F = \frac{m(v-u)}{t}$$

$$F_A = \frac{0.41(-10-20)}{0.1} = 123\ N$$

$$F_B = \frac{0.41(-10-20)}{0.6} = 20.5\ N$$

The ball applies almost $1/6^{th}$ the force on Goalkeeper B's gloves than it does on Goalkeeper A's hands. As we can see, there's a strong cushion effect for Goalkeeper B.

Challenge 8:

In the example above, the ball hits both hands at 50 m/s and both keepers punch the ball back at 10 m/s (the ball is 10 m/s after the punch). The ball contacts Goalkeeper A's hands for 0.05 seconds while it is in contact with Goalkeeper B's gloves for 0.5 seconds. What is the force applied on Goalkeeper A's hands and Goalkeeper B's gloves? Complete the free body diagram below and calculate.

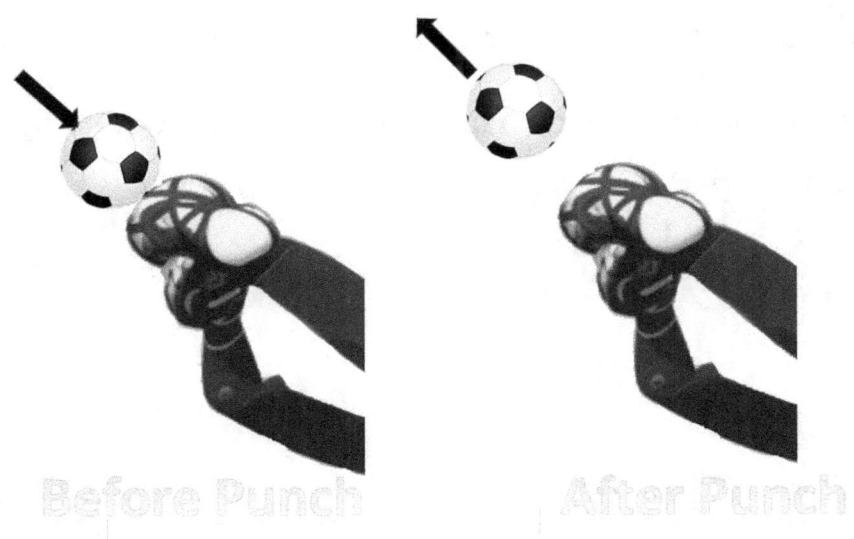

Angled Shots: The More You See, the More you Score.

Young soccer players learn about the geometry of the game even before they enter the classroom. They learn that if they move the ball towards the center they get a better angle for the shot, and score more often.

Young defenders also learn that if they force the striker wide before shooting, they have a better chance of saving the goal.

Goalkeepers also learn to move forward to reduce the angle that the striker can see.

Let's have a look at an example below:

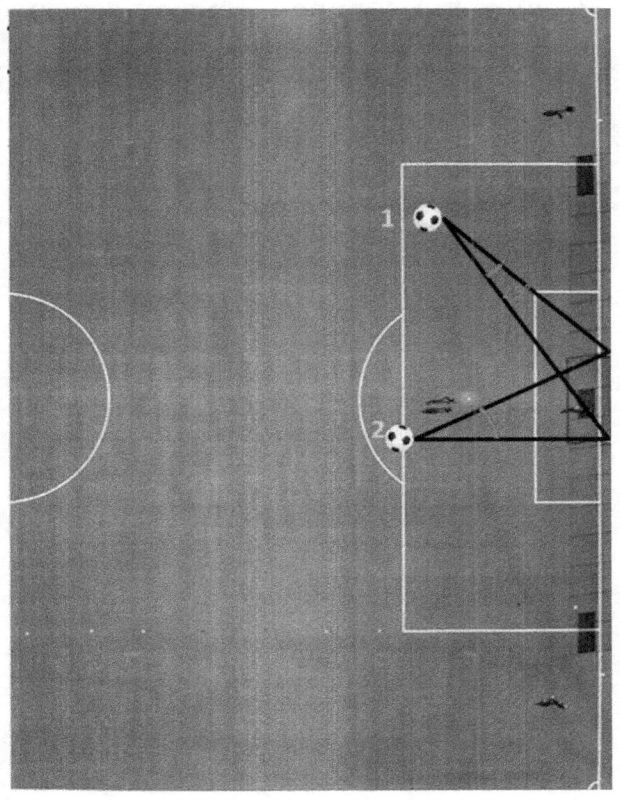

In diagram above, we look at two different positions marked 1 and 2. Position 1 has a 25 degrees angle view of the goal. position 2 has a 47 degree angle view of the goal.

Which position is better for scoring? Position 2 is much better than Position 1, assuming the player can shoot with both feet equally. The higher the angle, the better the position.

Sample Exercise:

Challenge 9:

Which of the below spots (1,2,3) are most likely to result in a goal?

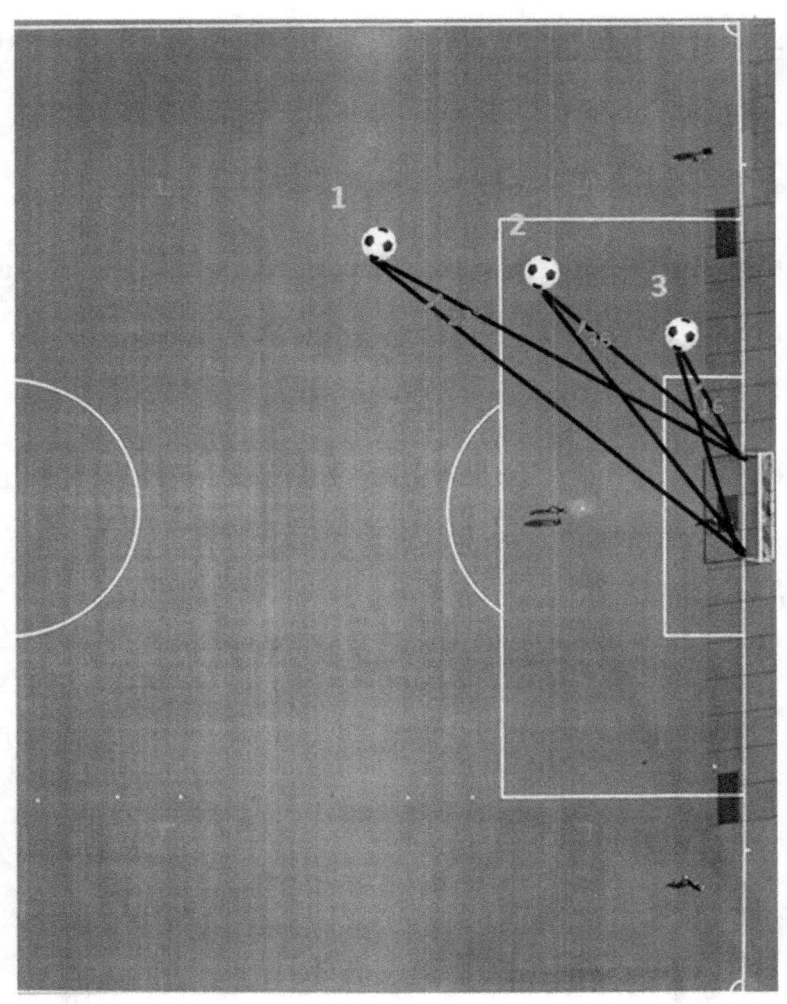

So, is there actually a way to calculate the angle? Yes, there is.

It uses Pythagros theorem below:

The Pythagros theorem states that in a right-angled triangle, the sum of the squares of the two sides of the right angle is equal to the square of the hypotenuse.

In the example below, **h** is the hypotenuse, **a** and **b** are the two sides of the right angle. Θ is the angle between the hypotenuse and side **b**

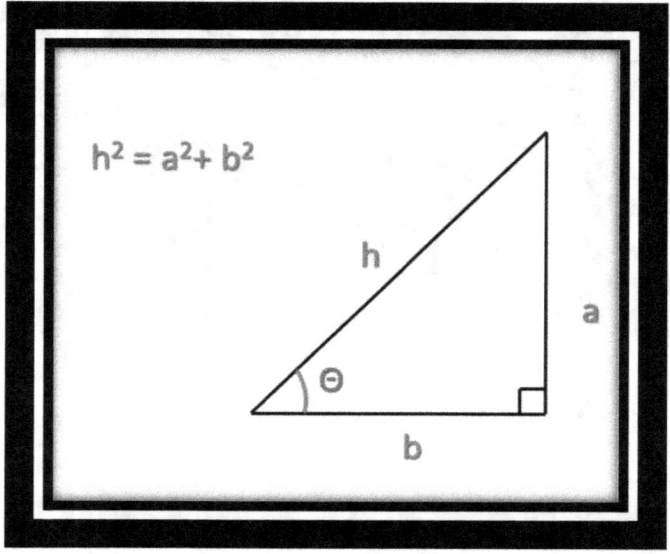

$$\sin \Theta = \frac{a}{h}$$

$$\cos \Theta = \frac{a}{h}$$

$$\tan \Theta = \frac{a}{h}$$

So, how does this work on the soccer field? Let's look at the diagram below.

The point where player contacts the ball forms two right angles with each post.

The orange triangle forms a right angle with the near post.

The blue triangle forms a right angle with the far post.

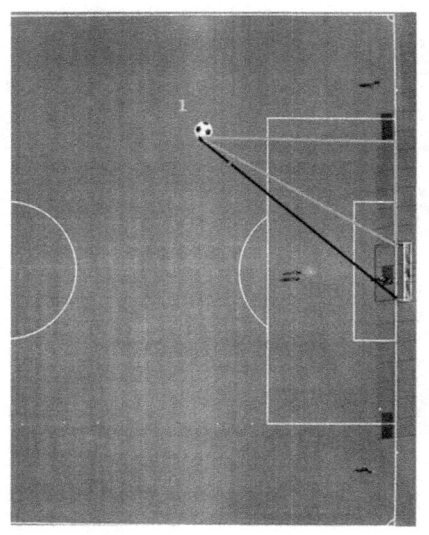 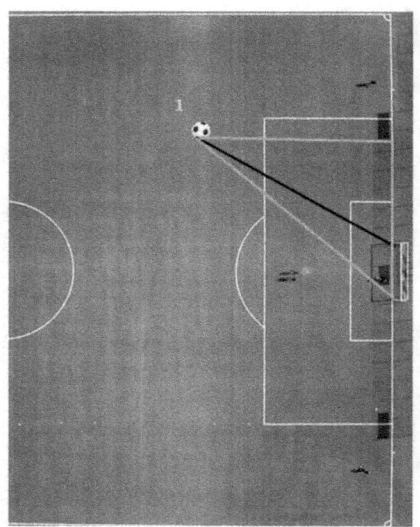

Now, let's say the spot is 25 m (75 ft) away horizontally and 15 m (45 m) away vertically from the near post.

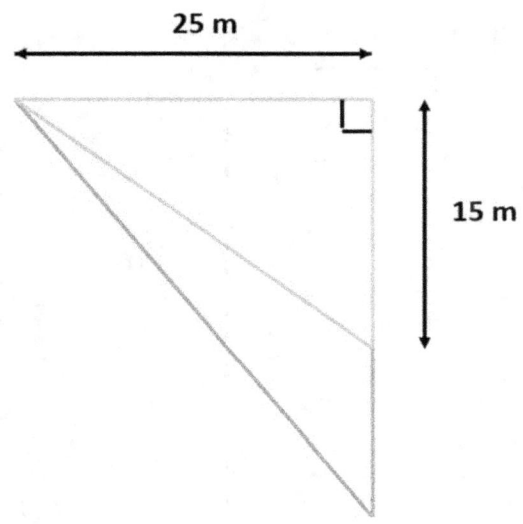

Now, if the goalpost is 7 m wide (or 21 feet wide), we get:

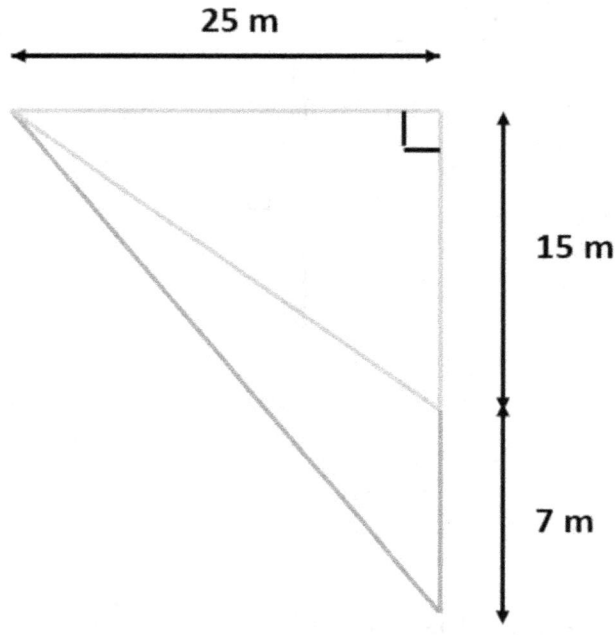

So, to solve this, we break it down into two triangles and find each angle.

First, let's do the orange triangle. In the orange triangle Θ^1 is the angle between the hypotenuse and the horizontal distance. We get Θ^1 as 30.96°.

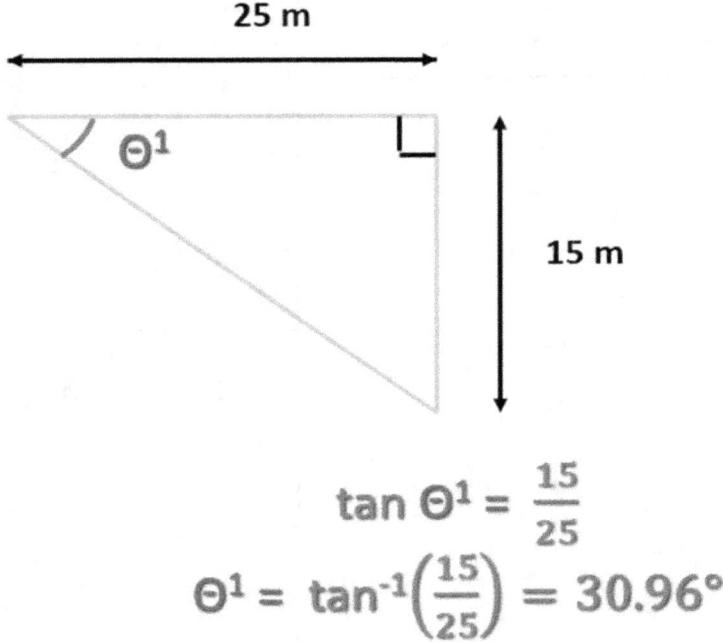

$$\tan \Theta^1 = \frac{15}{25}$$

$$\Theta^1 = \tan^{-1}\left(\frac{15}{25}\right) = 30.96°$$

Now, let's do the blue triangle. In the blue triangle Θ^2 is the angle between the hypotenuse and the horizontal distance. We get Θ^2 as 41.37°.

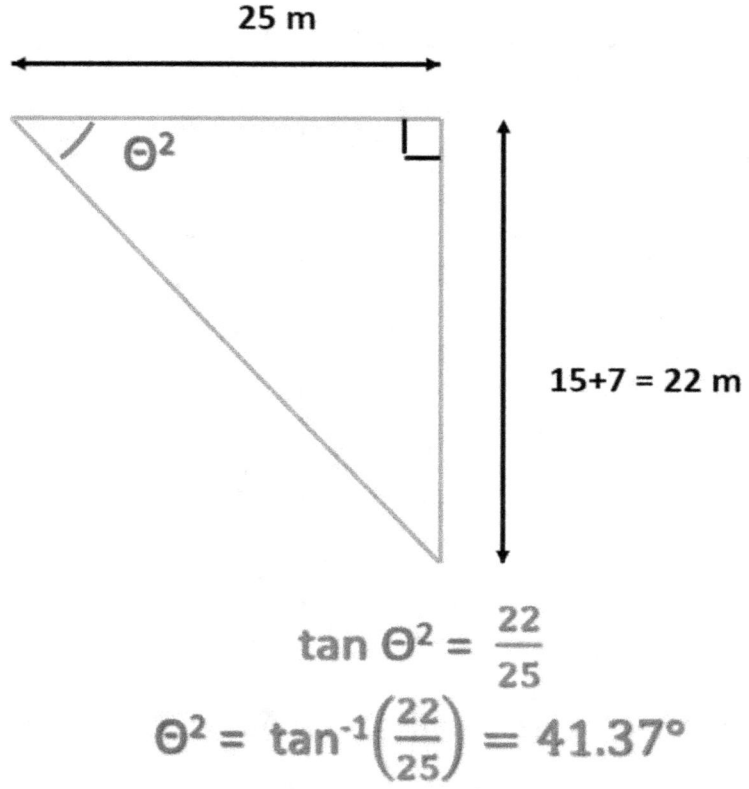

$$\tan \Theta^2 = \frac{22}{25}$$

$$\Theta^2 = \tan^{-1}\left(\frac{22}{25}\right) = 41.37°$$

Θ¹ and Θ² are shown in the diagram below.

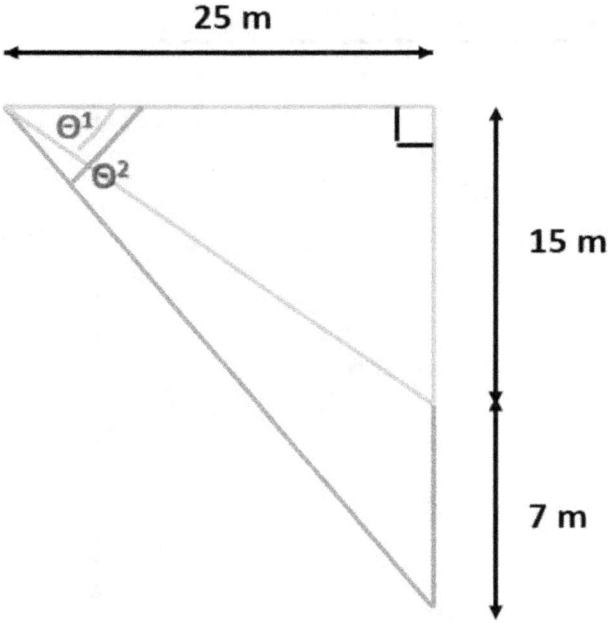

The difference between the two angles ($\Theta^2 - \Theta^1$) gives you the actual value of the angle between the posts.

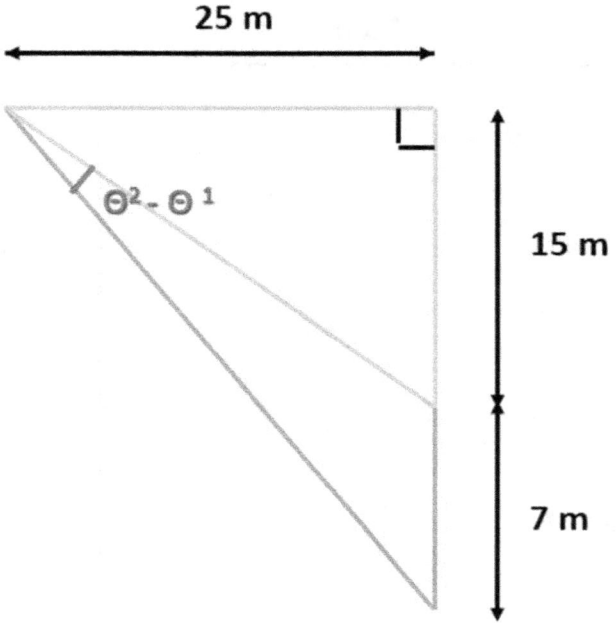

$\Theta^2 - \Theta^1$ is equal to $(41.37° - 30.96° = 11.41°)$

So, the angle between the two goal posts is **11.41°** for a point that's 25 m away horizontally and 15 m away vertically.

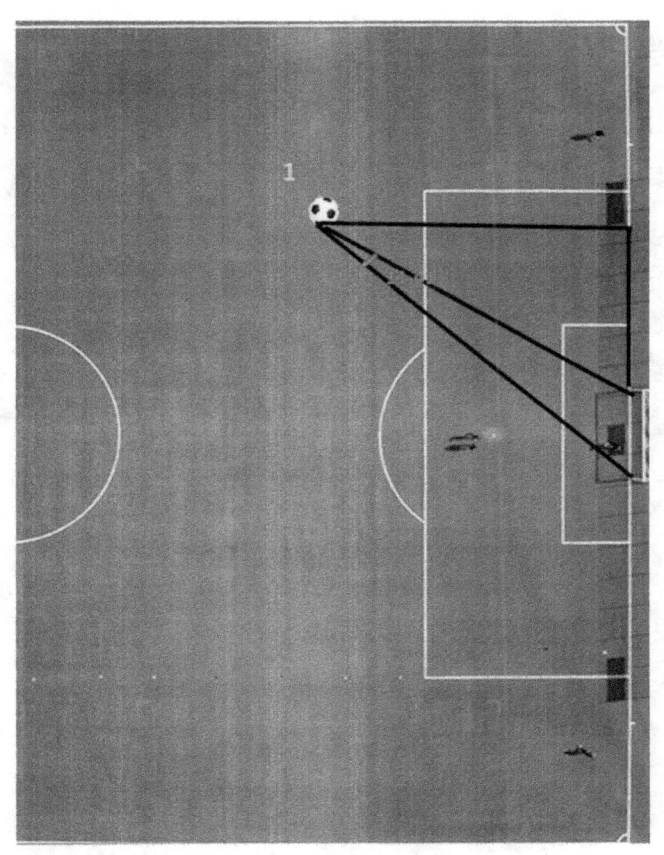

Challenge 10:

What is the angle that a spot makes between the goalposts? The spot is 15 m away horizontally and 15 m away vertically as shown below.

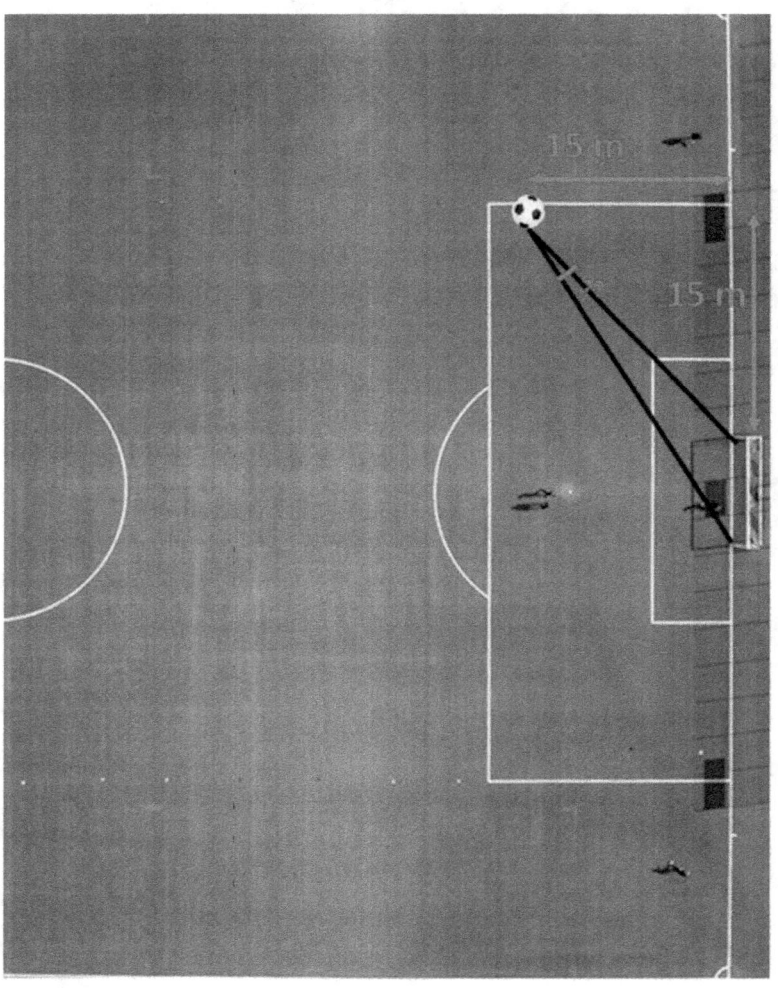

You can do your rough workings on sheet below…

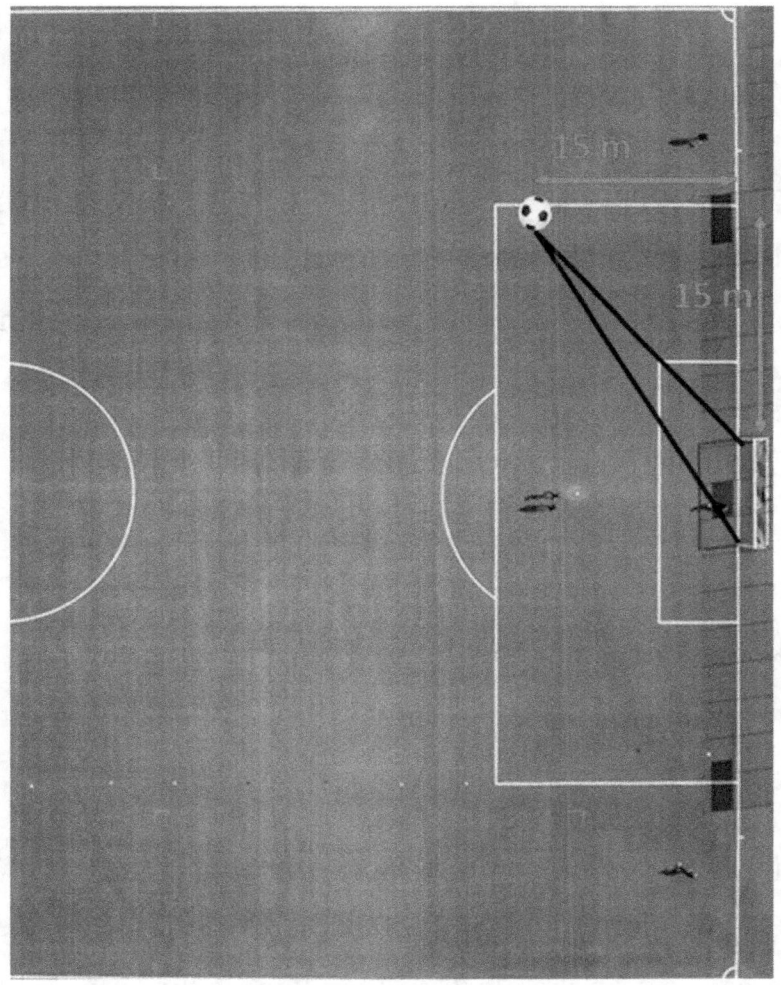

How the Keeper Fits Into This

You might have noticed plenty of times when the keeper has stepped out to get closer to the striker. This might strike you as a little strange. Doesn't the keeper have more time to see the ball and react if he stays backwards? Well, it's a bit more complicated than that.

You have just learnt how to calculate the shooting angle between the goal posts and the ball. However, one thing that calculation doesn't take into account is the goalkeeper. The goalkeeper is a small spot in the goal if he stays put. However, if the goalkeeper moves forward, he cuts into the angle a bit more; and the striker sees less of the ball.

The picture below explains this concept a bit more. The red lines are the portion of the goal that is blocked by the keeper. The area between the black and red lines is the area that the striker can score.

As we can see, when the keeper moves forward, the area available to score is much smaller, and this reduces the chances of a goal happening.

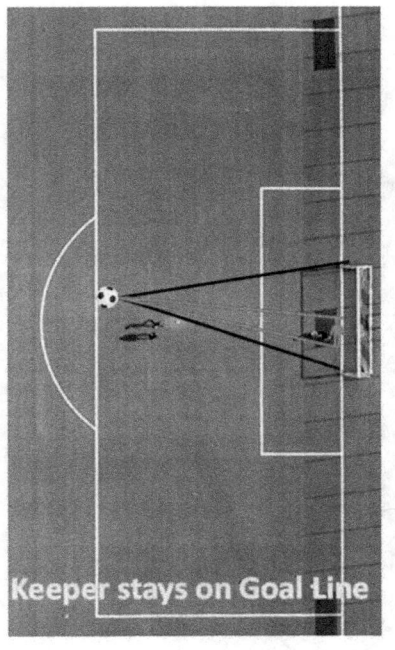 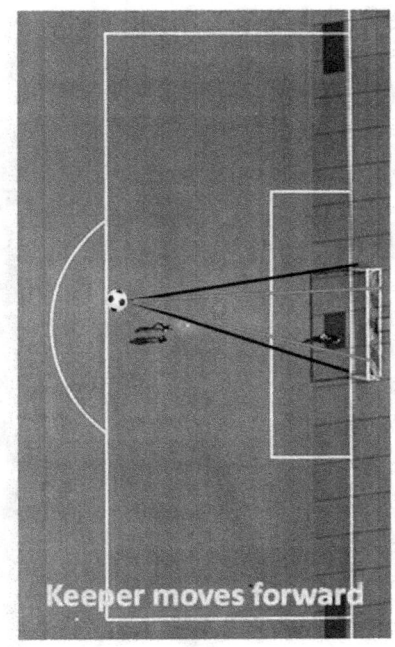

Also, the wider the keeper is able to stretch while moving forward, the more of the goal he can block. That's why we often see keepers stretch as wide as possible while facing a striker.

Practice Exercises

For the below diagrams, draw straight lines that show the viewing area of the goal; and the viewing area blocked by the goalkeeper.

Challenge 11:

Challenge 12:

Challenge 13:

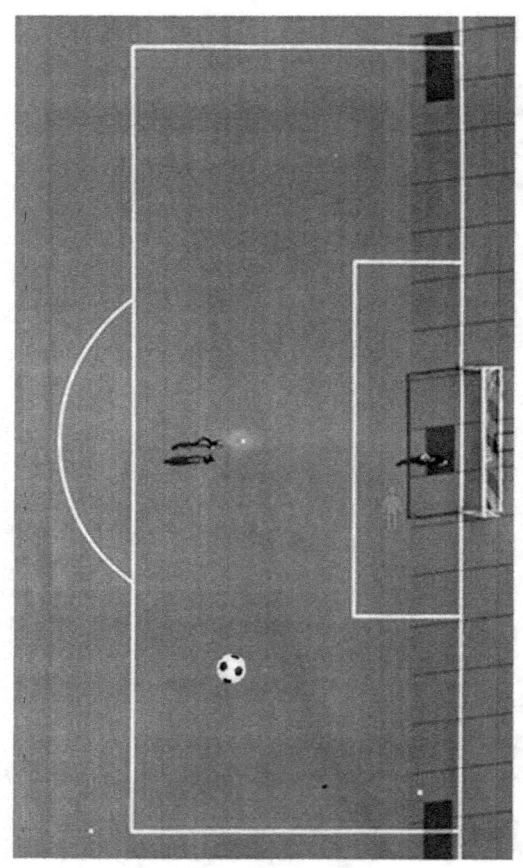

Bonus: What a Drag

What is Drag?

Drag is the air resistance experienced by the ball while it is moving through the air. It's like an extra force experienced by the ball in a direction opposite to the direction of motion.

This causes the ball to travel a shorter distance and move to a lower height.

Drag on a Soccer Ball

When a soccer ball moves through the air, the air separates along the surface of the soccer ball and moves over the back of the ball as shown below.

Behind the soccer ball, there is an area of turbulence called the wake of the flow. The wake of the soccer ball causes the drag force.

Soccer Ball Wake

Reduce Drag on Soccer Ball

One way to reduce drag on a soccer ball is to reduce the size of the wake behind the soccer ball. One strategy is to add more panels to the soccer ball, and this causes the wind to stick closer to the surface of the ball as it moves over. It reduces the size of the wake behind as shown below. Another way to increase this is to add dimples or find other ways to increase surface roughness.

This is also the reason that a golf ball has dimples on the surface.

Conclusion

Thank you for taking the time to read this book.

I hope you have found it useful; and learnt a few basic engineering concepts. If you've been through the exercises, you now understand the basics.

If you liked my book and you'd like to learn more, check out my Author page below. I have half a dozen books out on coding for kids and beginners.

https://www.amazon.com/Bob-Mather/e/B07HHQZC4Y

If you like more books similar to this one, feel free to try these ones out below:

Fascinating Engineering Book for Kids: 500 Dynamic Facts

Awesome Engineering Activities for Kids

If you'd like any clarification regarding the topics; or any suggestions; please email me at: abiprod.pty.ltd@gmail.com

I also do one-one coaching online if you are interested in personalized help.

The end... almost!

Reviews are not easy to come by.

As an independent author with a tiny marketing budget, I rely on readers, like you, to leave a short review on Amazon.

Even if it's just a sentence or two!

So if you enjoyed the book, please browse to the product page and leave a review as shown below:

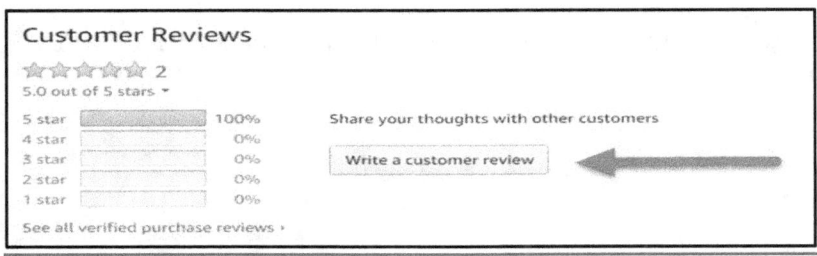

I am very appreciative for your review as it truly makes a difference.

Thank you from the bottom of my heart for purchasing this book and reading it to the end.

Challenge Answers

Challenge 1:

The mass of a soccer ball is 410 grams (0.41 kg). It is moving slowly, with an initial velocity of 2 m/s in the same direction of the kick.

The soccer player kicks a ball. His foot has an effective mass of 1 kg and his foot is moving at a velocity of 10 m/s.

The velocity of the foot after kicking is 7 m/s. The elasticity of the system is 0.92. Calculate the final velocity of the ball.

Challenge 1 Answer:

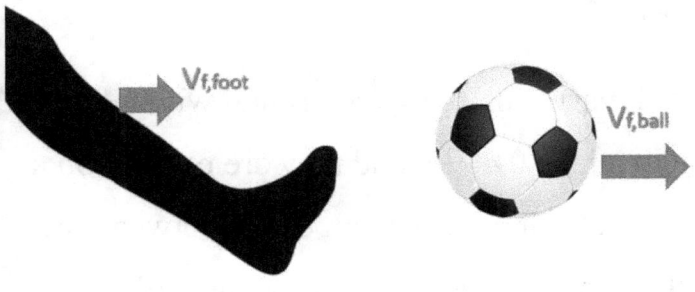

$m_{ball} = 0.41$ kg

$m_{foot, effective} = 1$ kg

$v_{i,ball} = 2$ m/s

$v_{i,foot} = 10$ m/s

$v_{f,ball} = ??$

$v_{f,foot} = 7$ m/s

$$(0.41 * 2 + 1 * 10) * 0.92 = (0.41 * v_{f,ball} + 1 * 7)$$

$$v_{f,ball} = 7.21 \; m/s$$

Challenge 2

Now, it's your turn.

Messi passes a stationary ball along the ground with a force of 100 N. The ball weighs 0.41 kg and they are playing on a soccer field with a coefficient of friction of 0.7 between the grass and the ball. What's the acceleration of the ball? Fill in the free body diagram below while solving the problem.

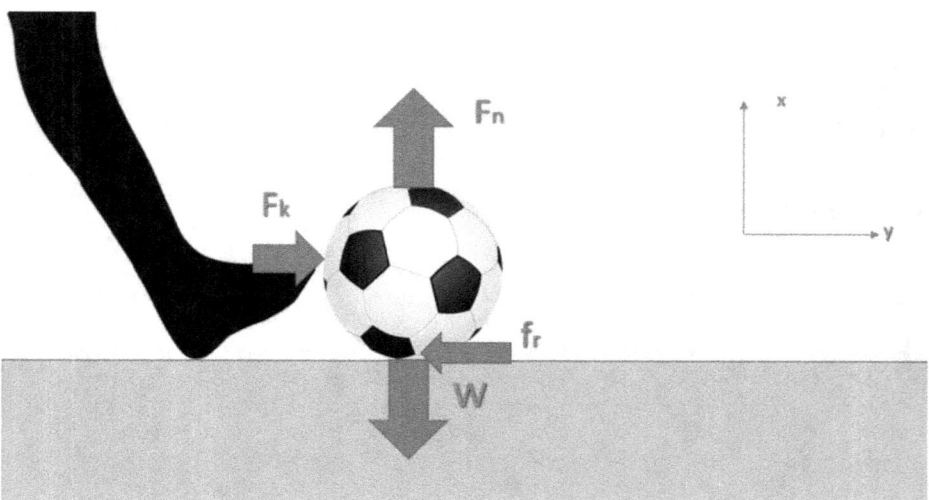

Challenge 2 – Answers

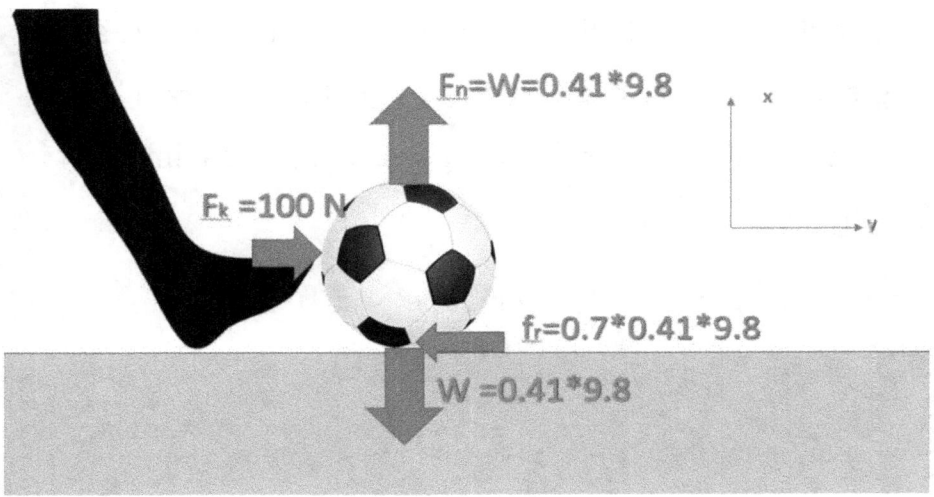

F_k = Force on the ball = 100 N

m = mass of the ball = 0.41 kg

μ = Coefficient of static friction = 0.7

g = acceleration due to gravity = 9.81 m/s²

$a_x = F_k/m - \mu g$

=100/0.41 – 0.7*9.8

=237 m/s²

Challenge 3

Tim kicks a soccer ball with an initial speed of 10 m/s at an angle of elevation of 45 degrees. Fill out the diagram below and figure out how fair the ball goes, and the maximum height it can achieve.

Bonus: Is it going to fly over a goalpost 10 m away?

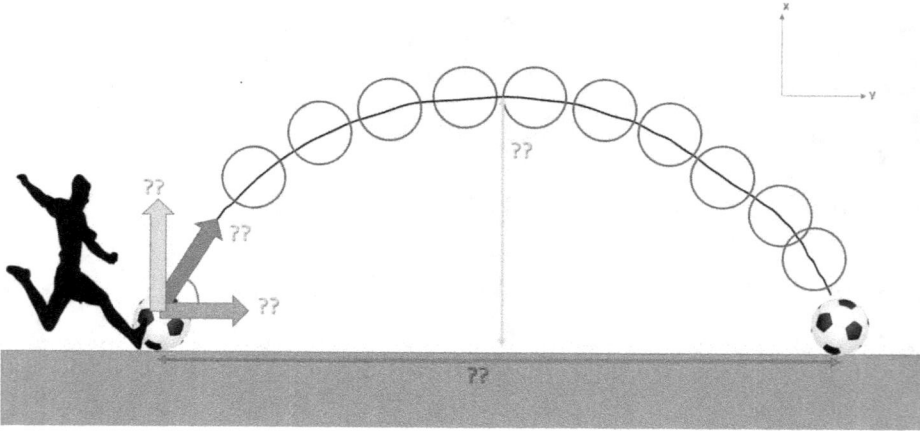

Challenge 3 – Answers

Based on the equation for maximum vertical height, we get:

$$H = \frac{u^2(\sin\theta)^2}{2g}$$

$$H = \frac{100(\sin 45)^2}{2*9.8} = 2.55\ m\ high$$

Based on the equation for horizontal range, we get:

$$R = \frac{u^2 \sin 2\theta}{g}$$

$$R = \frac{10^2 \sin 90}{9.8} = 10.2\ m$$

So, the ball travels a maximum height of 2.55 m before coming down. And it achieves a range of only 10.2 m.

So, it does not go over a goalpost 10 m away. The goalpost's height is 2.7 m but the maximum height the ball achieves in only 2.55 m.

Challenge 4

The ball is moving at a speed of 20 mph. The wind is blowing against the ball at a speed of 6 mph. What's the absolute and relative velocity of the ball?

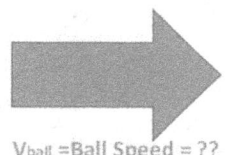

$V_{ball} =$ Ball Speed = ??

$V_{wind} =$ Wind Speed = ??

Draw Wind above

$V_{ball,rel} =$??
$V_{ball,abs} =$??

Challenge 4 – Answers

$V_{ball} =$ Ball Speed = 20 mph

$V_{wind} =$ Wind Speed = 6 mph

$V_{ball,rel} = 20 - 6 = 14$ mph
$V_{ball,abs} = 20$ mph

Challenge 5

A ball is moving at a velocity of 37 mph in the direction of wind. The wind is blowing at 10 mph. What is the absolute velocity and relative velocity of the ball?

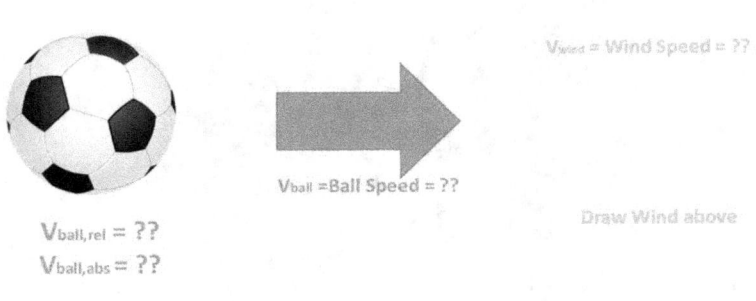

$V_{ball,rel}$ = ??
$V_{ball,abs}$ = ??

V_{ball} = Ball Speed = ??

V_{wind} = Wind Speed = ??

Draw Wind above

Challenge 5 – Answers

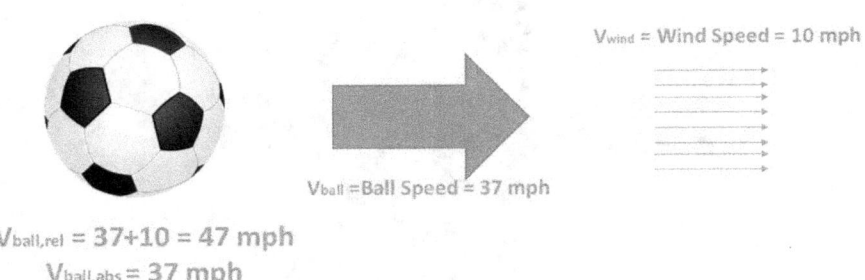

$V_{ball,rel}$ = 37+10 = 47 mph
$V_{ball,abs}$ = 37 mph

V_{ball} = Ball Speed = 37 mph

V_{wind} = Wind Speed = 10 mph

Challenge 6

Here's the ball with the wind flowing against it. In the empty sheet below, draw the areas of high and low pressure and the direction it is likely to bend.

Challenge 6 – Answers

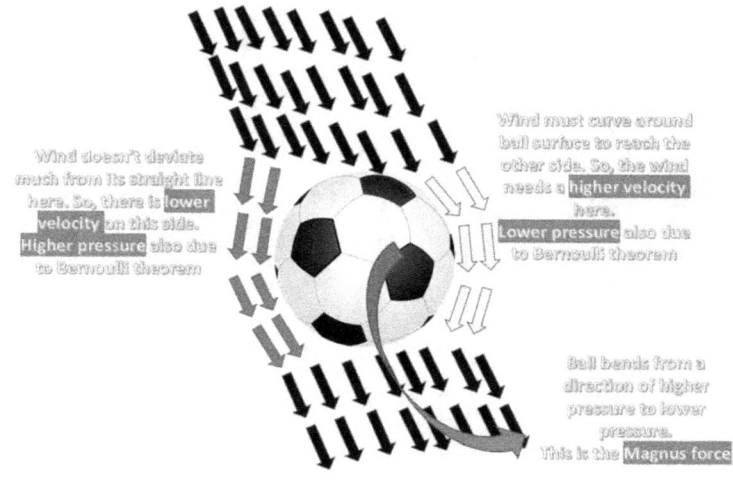

Challenge 7:

Challenge 7 – Answers

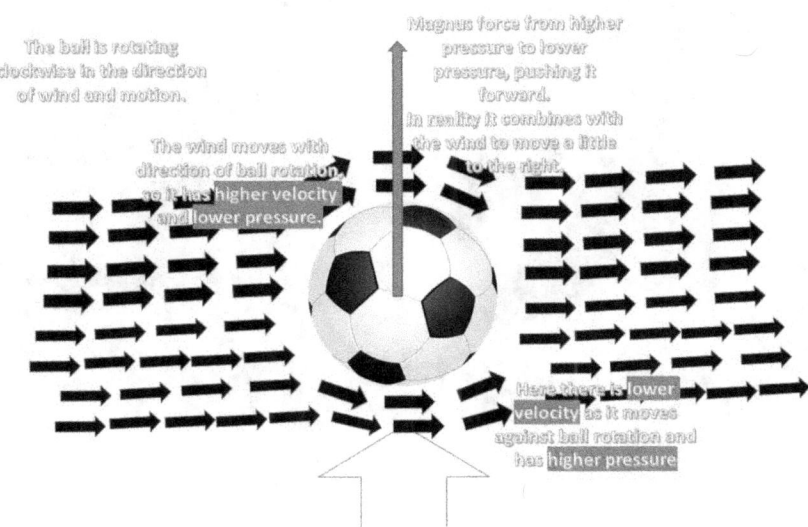

Challenge 8:

In the example above, the ball hits both hands at 50 m/s and both keepers punch the ball back at 10 m/s (the ball is 10 m/s after the punch). The ball contacts Goalkeeper A's hands for 0.05 seconds while it is in contact with Goalkeeper B's gloves for 0.5 seconds. What is the force applied on Goalkeeper A's hands and Goalkeeper B's gloves? Complete the free body diagram below and calculate.

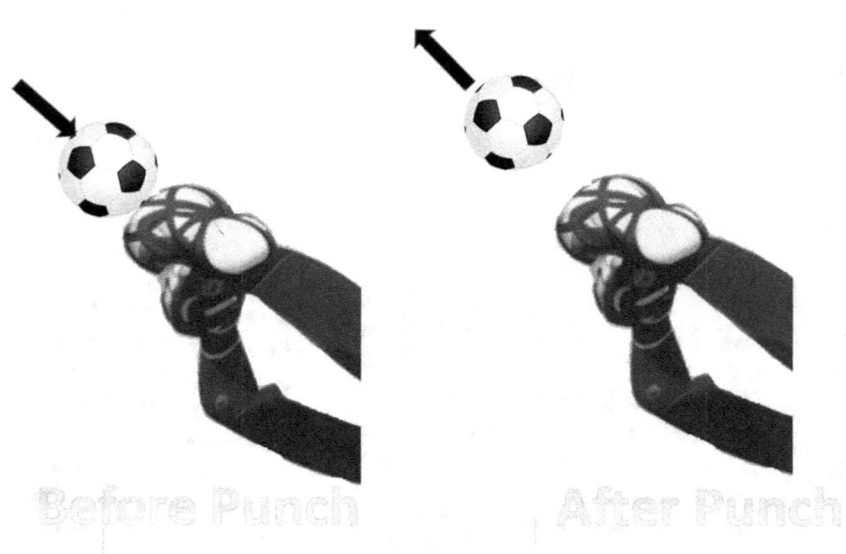

Challenge 8 – Answers:

So, we have:

$m = 0.41$ kg

$u = 50$ m/s

$v = -10$ m/s (it goes in opposite direction so -ve sign)

$$F = \frac{m(v-u)}{t}$$

$$F_A = \frac{0.41(-10-50)}{0.05} = 492 \text{ N}$$

$$F_B = \frac{0.41(-10-50)}{0.5} = 49.2 \text{ N}$$

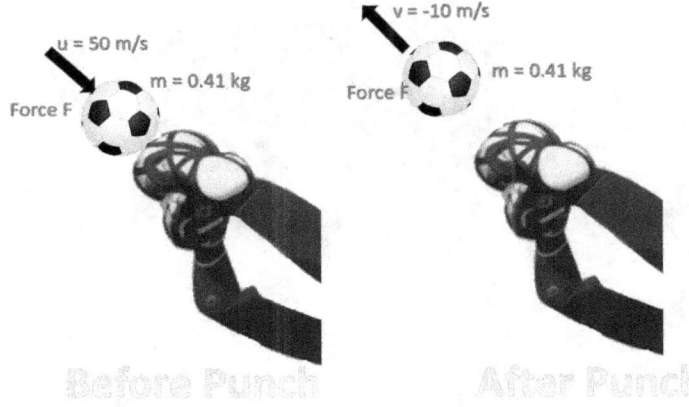

Challenge 9:

Spot Number 2 has the highest chance of a goal.

Challenge 10:

What is the angle that a spot makes between the goalposts. The spot is 15 m away horizontally and 15 m away vertically as shown below.

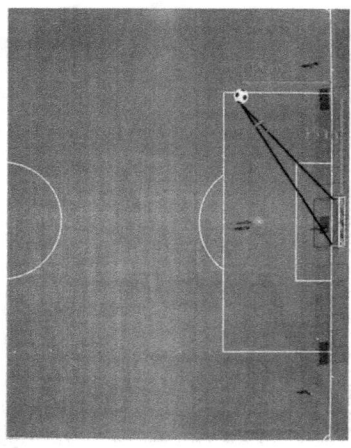

Challenge 10 - Answers:

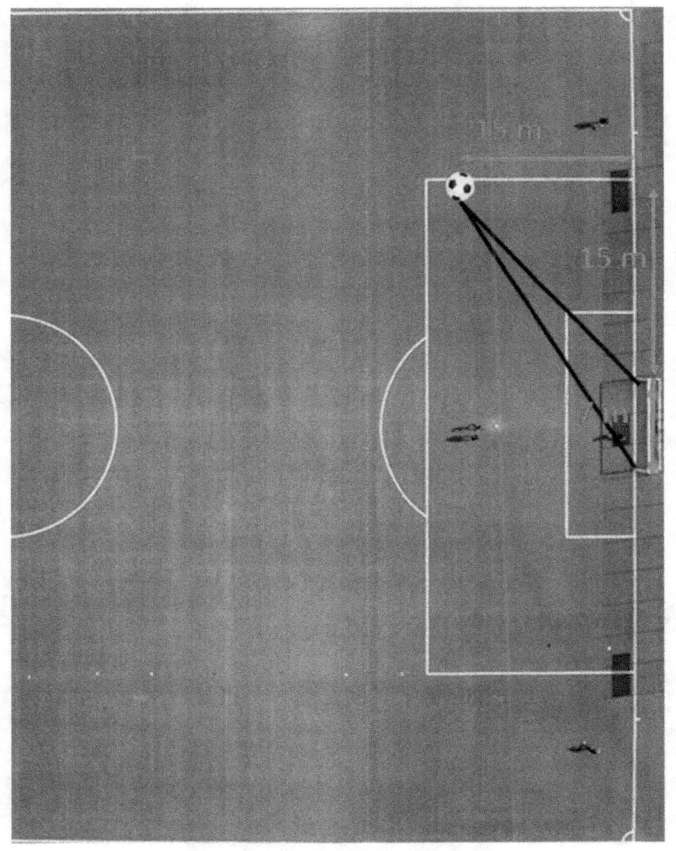

$$\tan^{-1}\frac{22}{15} - \tan^{-1}\frac{15}{15} = \mathbf{10.71°}$$

Challenge 11:

Challenge 11 - Answers

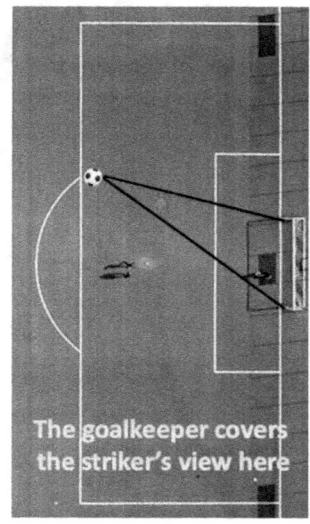

The goalkeeper covers the striker's view here

Challenge 12:

Challenge 12 - Answers

Challenge 13:

Challenge 13 - Answers:

Coding & Logic STEM Activity Book for Kids

Table of Contents

Table of Contents 108
Disclaimer ....**Error! Bookmark not defined.**
Introduction 109
The Search for that Elusive Banana 112
Taxi Please 129
Number Key Challenge 149
Loops 163
Nested Loops 173
Conditional Loops 181
Programming Challenges 186
Answers 195
The Search for that Elusive Banana 196
Taxi Please 210
Number Key Challenge - Answers 220
Loops 223
Nested Loops 228
Conditional Loops 232
Programming Challenges 235

Free Gift

We do want you to succeed in coding. To ensure your success, we are giving you a free list of projects that you can work on once you are completed with this book.

https://coding.gr8.com/

Introduction

It was a lovely day in Miami. Sam was enjoying the baseball game while working from home. Then he heard a beep, after which he was staring at a blue screen of death. The worst possible time this could happen. Sam was despondent and didn't feel like taking it to the repair shop. Sam yelled loudly in frustration. On hearing this, his 8-year-old son asked what happened. On learning about the blue screen, his son told him not to worry. He took the laptop and started diagnosing the problem. He checked for any hardware issues and verified that there were no issues there. He restarted the computer and did a complete software scan. He uninstalled a malfunctioning software that Sam had installed a day earlier. Within 30 minutes, Sam's computer was as good as new. What Sam's son had just demonstrated was logical thinking and step-by-step troubleshooting. These are essential skills for programming as well. These are the skills that you will learn in this book.

This workbook has activities designed to improve your logic and problem-solving abilities! It's the best way to start coding. It's filled with basic exercises that are designed for kids to improve creativity, problem solving, logic and critical thinking. These are more skills that are essential for programming.

It contains a diverse range of exercises, from beginner level scratch arrow programs to loop prediction exercises. It starts off easy to increase engagement and increases in difficulty as one progresses through the book. It unravels the mysteries of coding one step at a time. It allows one to focus on the logic and problem-solving aspect of programming.

While this workbook has been designed for young kids, it can be used by anyone who's struggling to understand the basics of coding.

This workbook is self-paced, and one can progress at a speed that suits then. One can go through this book independently or with the help of a mentor. It's a great place to experiment and learn from them. You ready to get started?

The Search for that Elusive Banana

Below is a series of challenges that are great for those who have never done any coding before, especially young kids. It might seem simple at first, but it's the first basic step to the process of sequential thinking. Sequential thinking forms the basis of programming, where you need to break a problem down into individual parts and solve them one step at a time.

In the challenges below, use arrows below to guide the monkey to the target banana using the shortest path. A lot of these have two possible paths or solutions. Answers to all problems are at the back of the book.

Arrows That Can be Used:

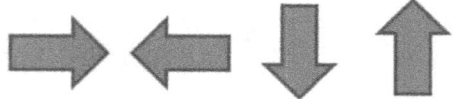

Use the first challenge below and the answer to understand the puzzle.

Challenge 1

Challenge 1 – Answers

Challenge 2

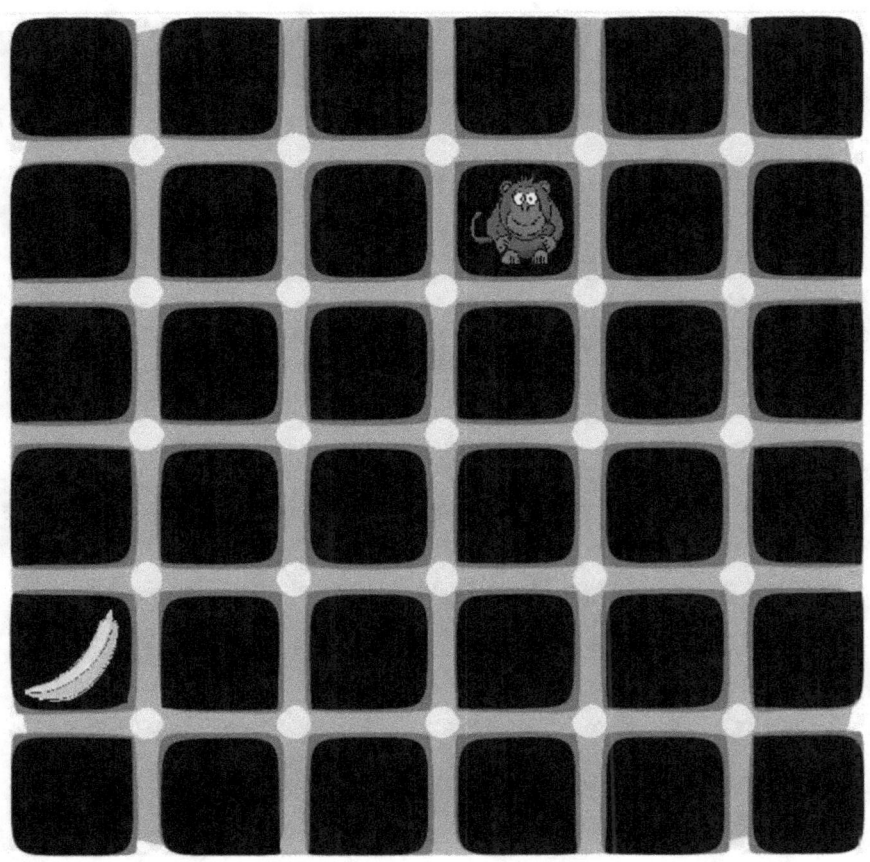

Challenge 3

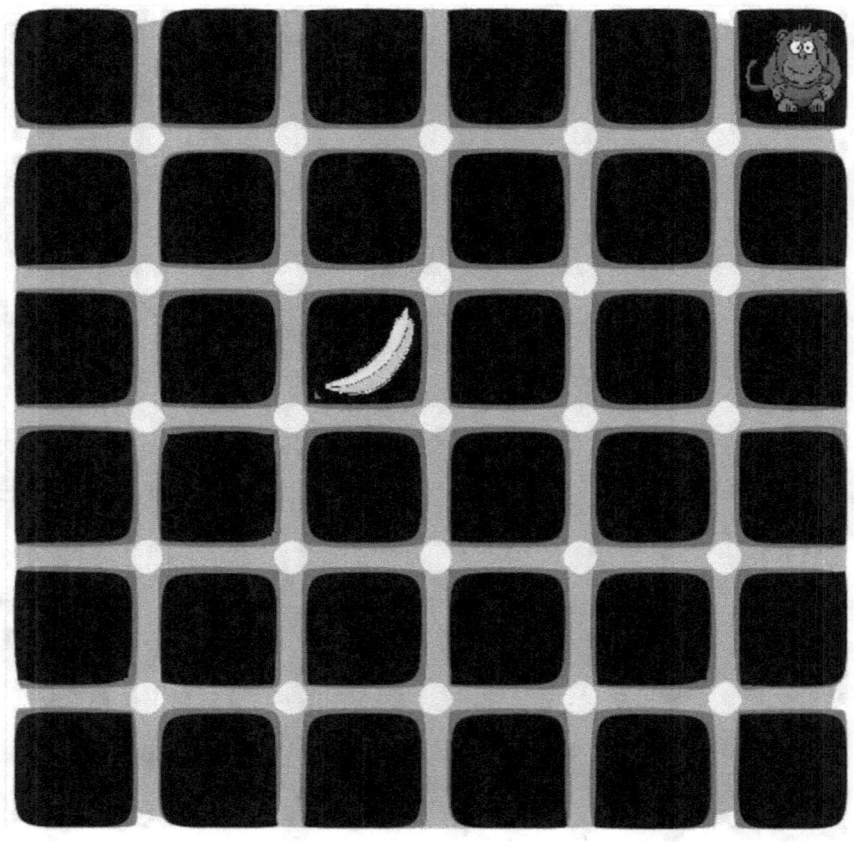

Challenge 4

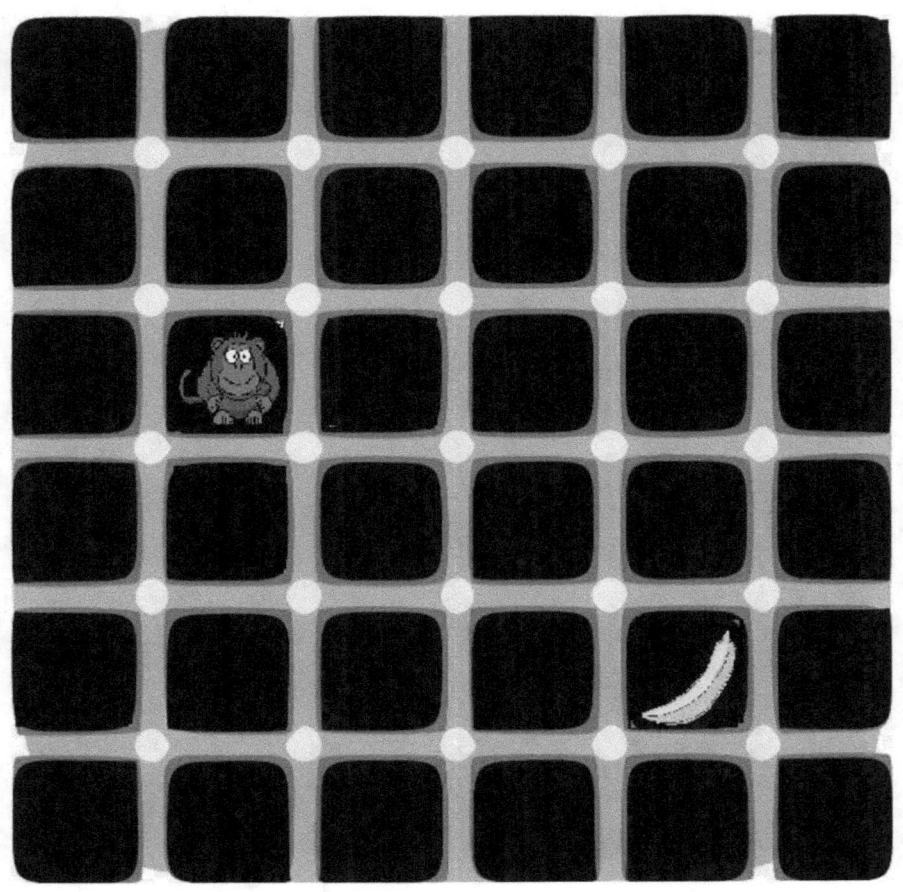

Challenge 5

Challenge 6

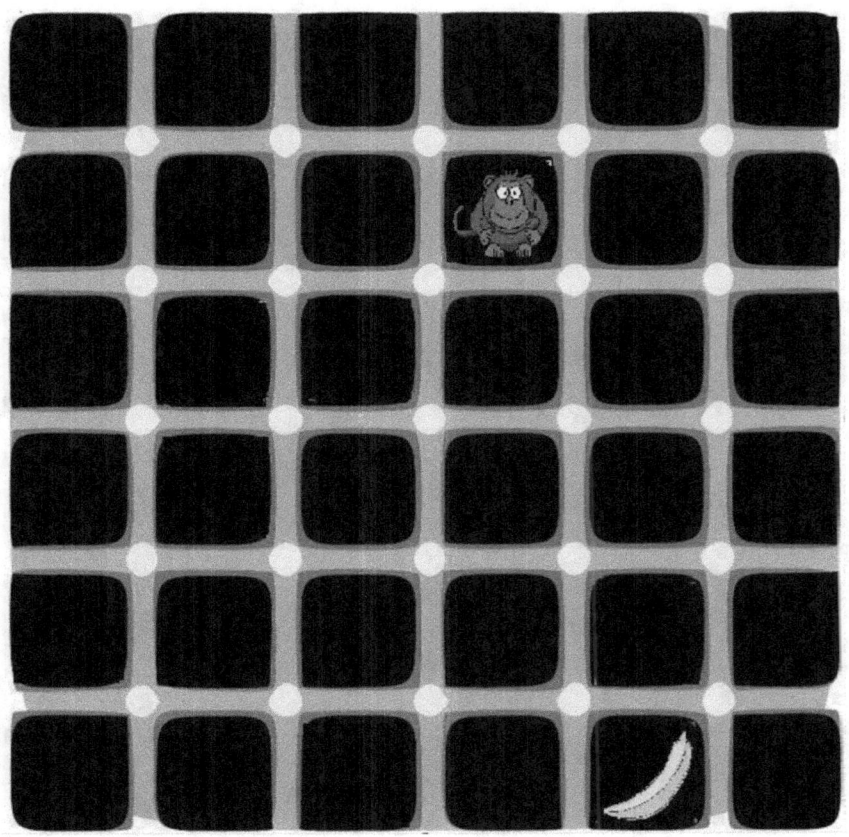

Challenge 7

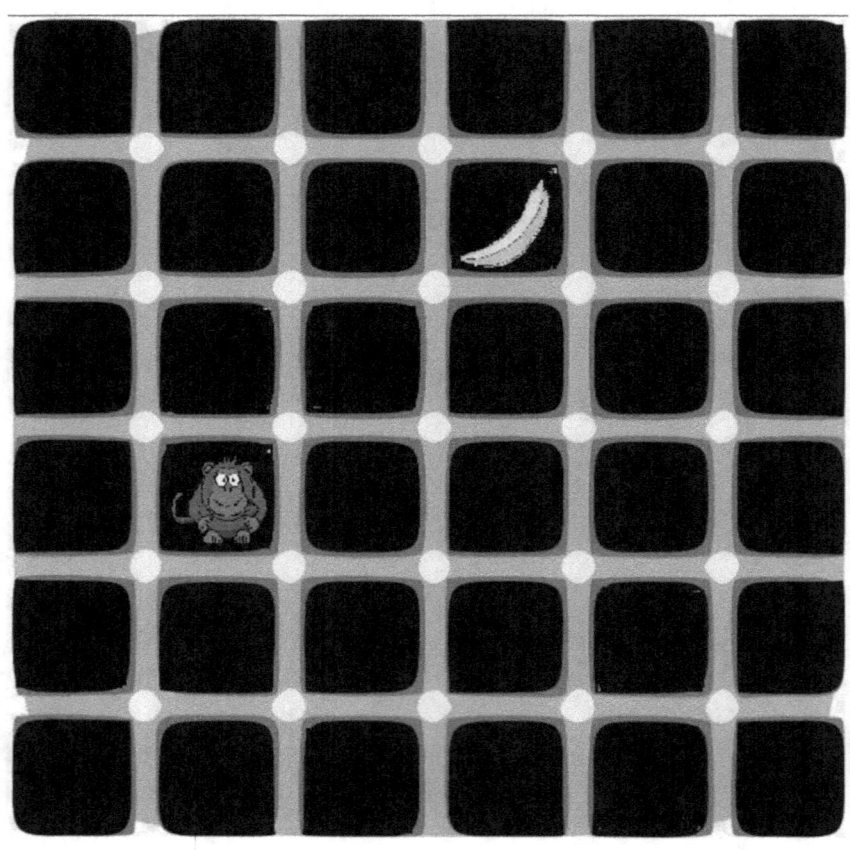

Challenge 8

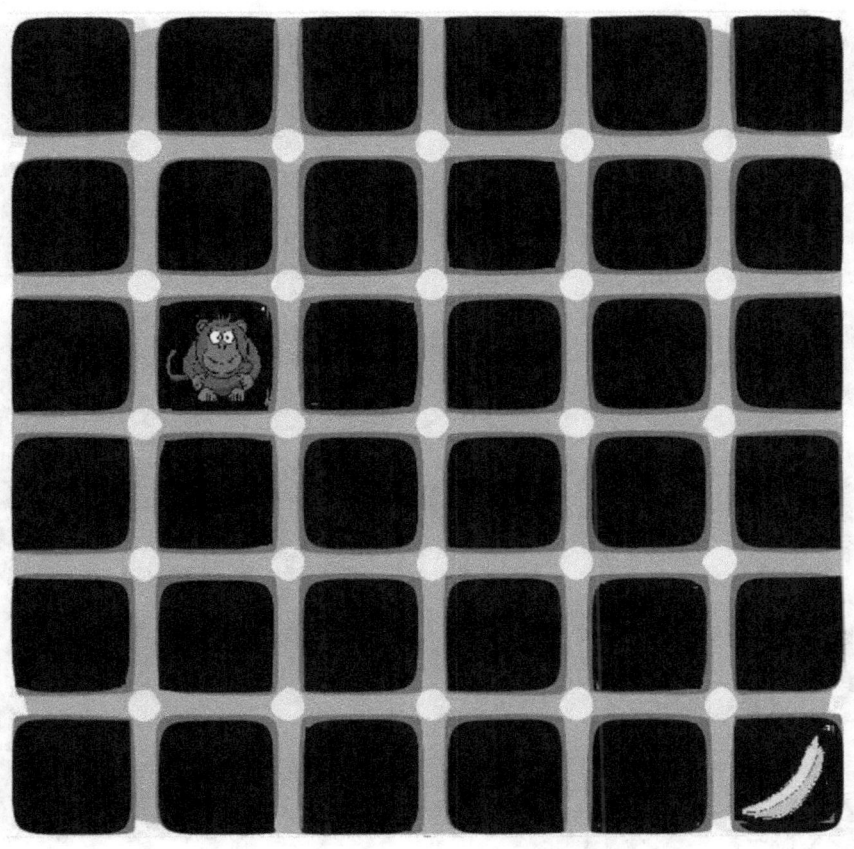

Challenge 9

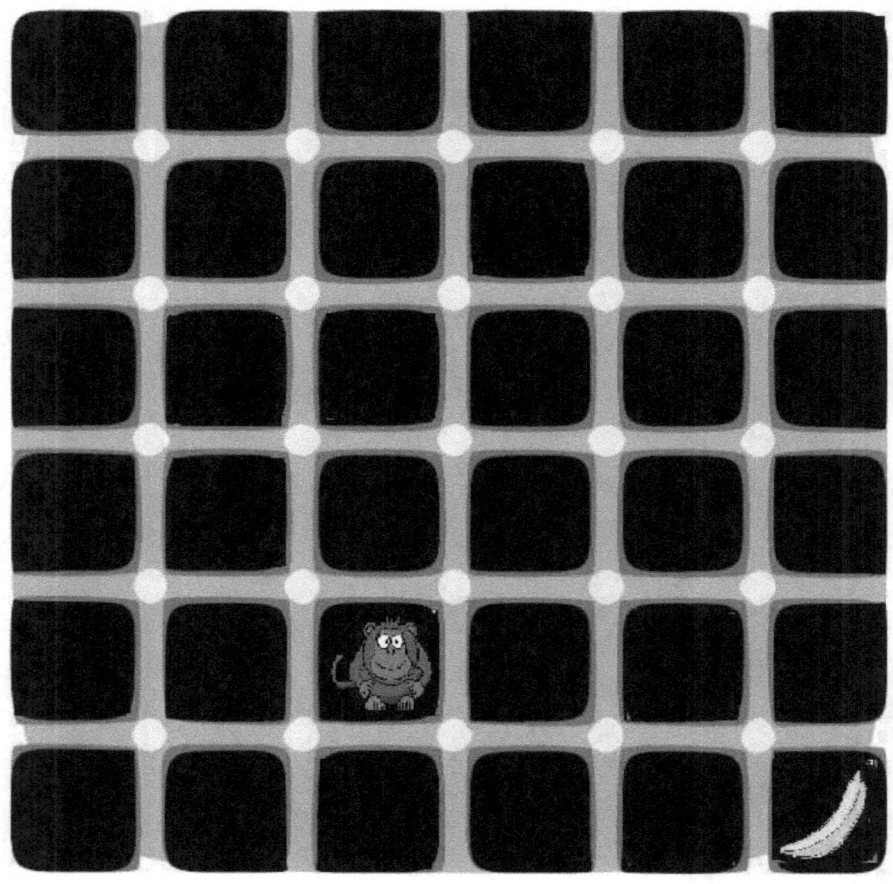

Challenge 10

Now, we have some obstacles in the monkey's path. Find the shortest distance to the banana while avoiding the rocks.

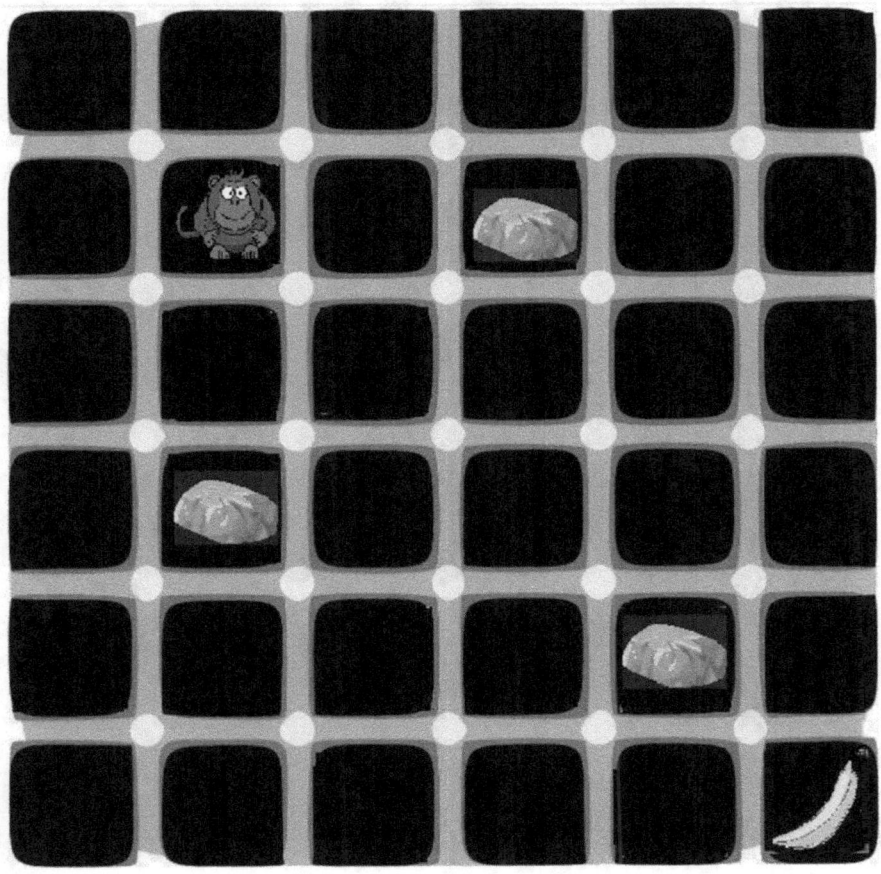

Challenge 11

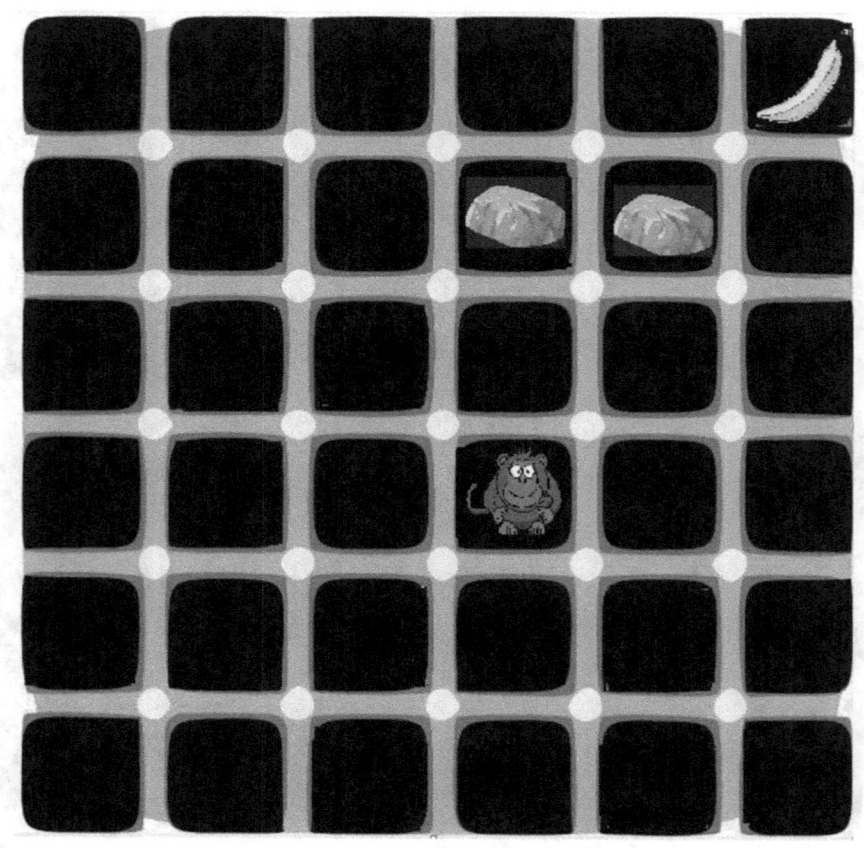

Challenge 12

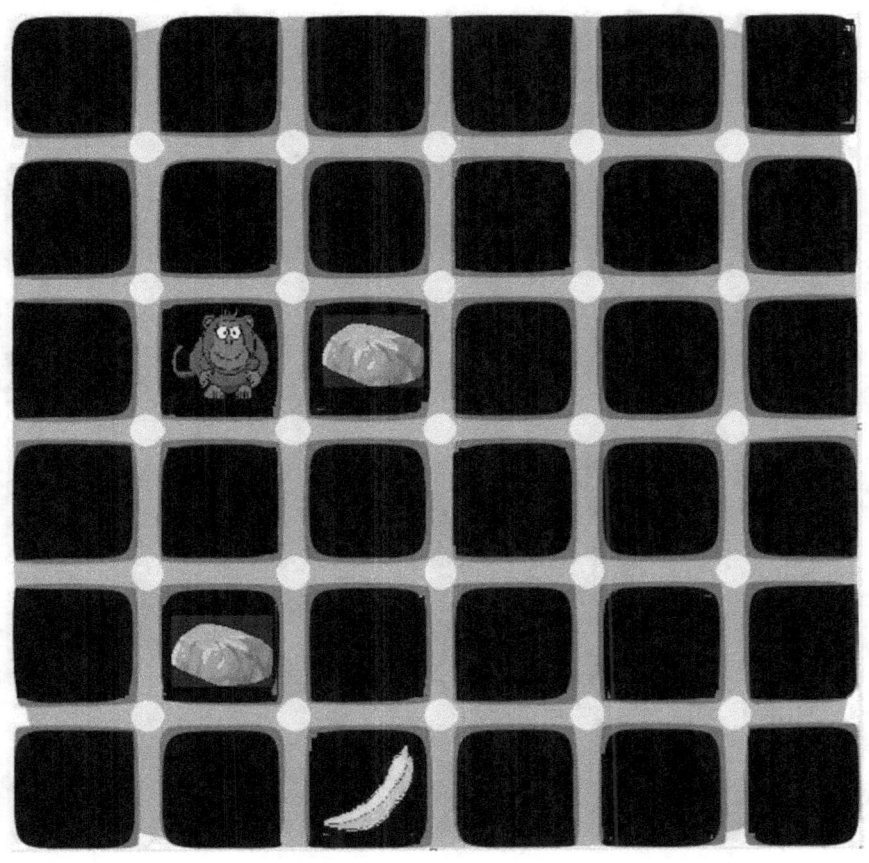

Challenge 13

Challenge 14

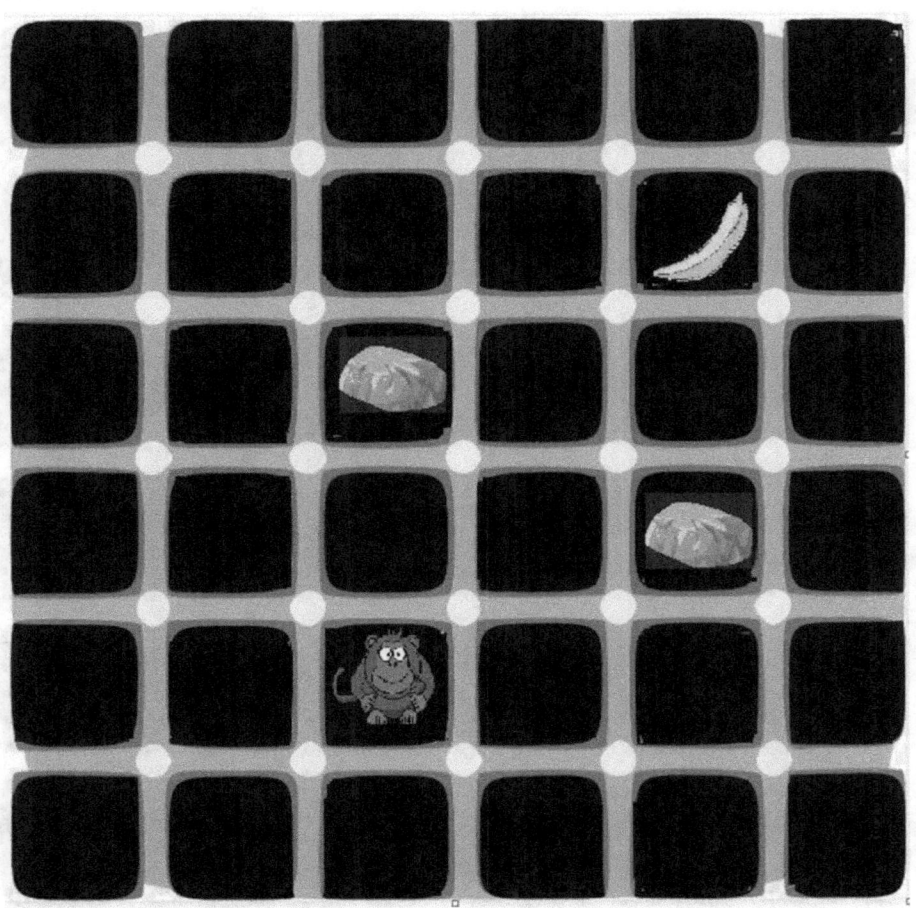

Challenge 15

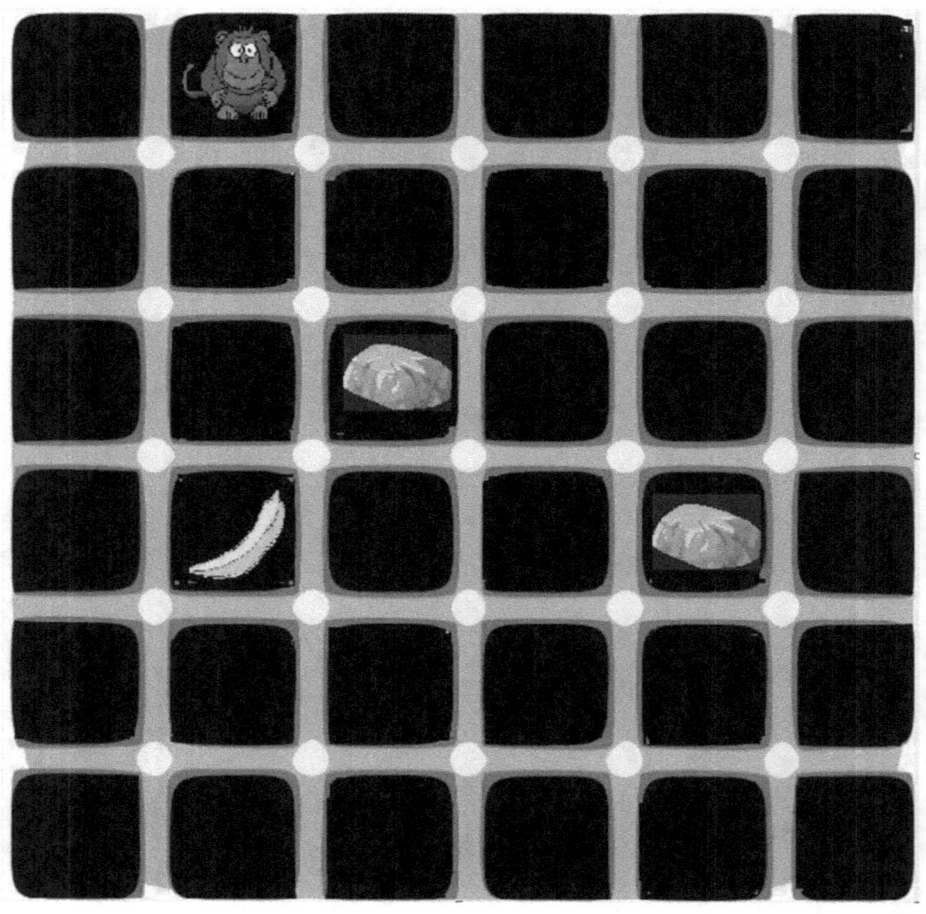

Taxi Please

In the next few challenges, the taxi needs to pick up the woman and give her a ride to her home. Use arrows in the same way as was done for previous challenges to find the shortest path.

Challenge 16

Challenge 17

Challenge 18

Challenge 19

Challenge 20

Challenge 21

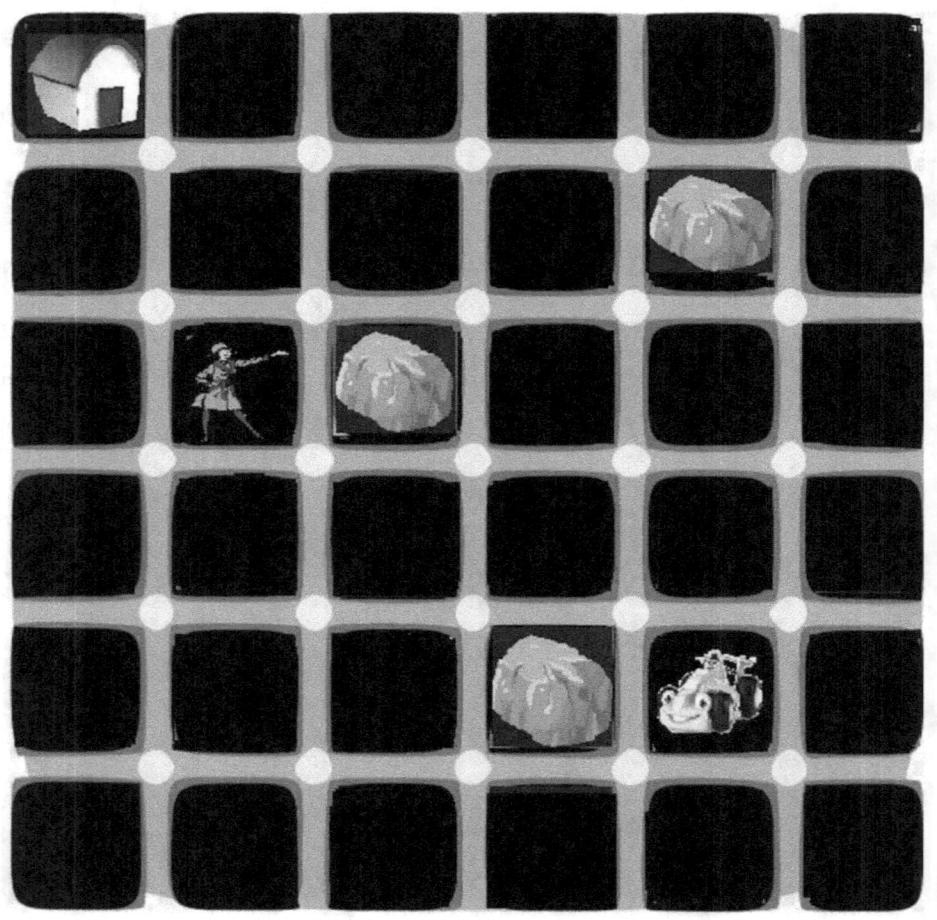

Challenge 22

Challenge 23

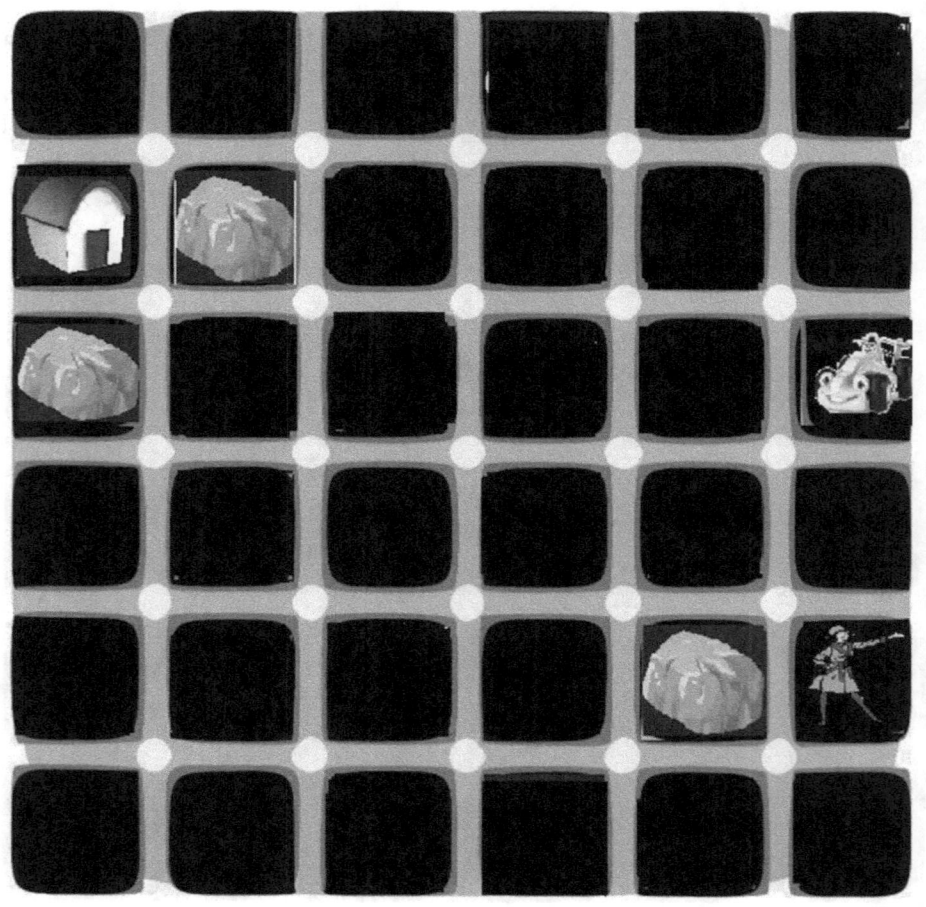

Challenge 24

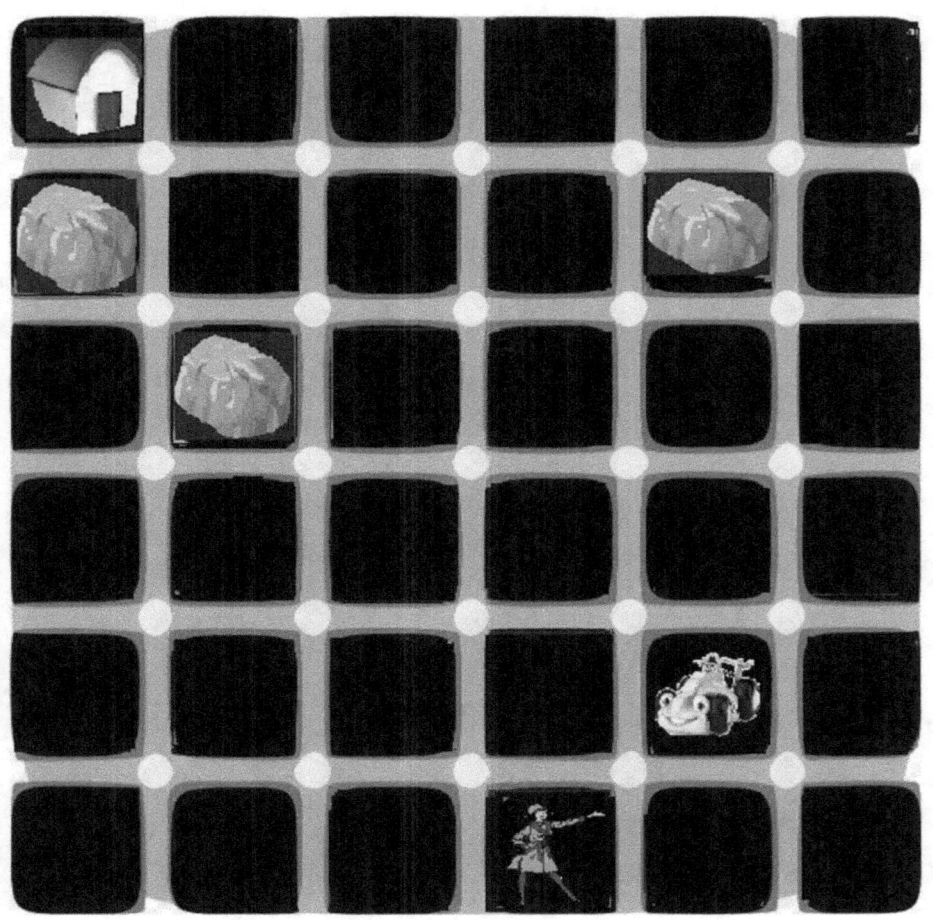

Challenge 25

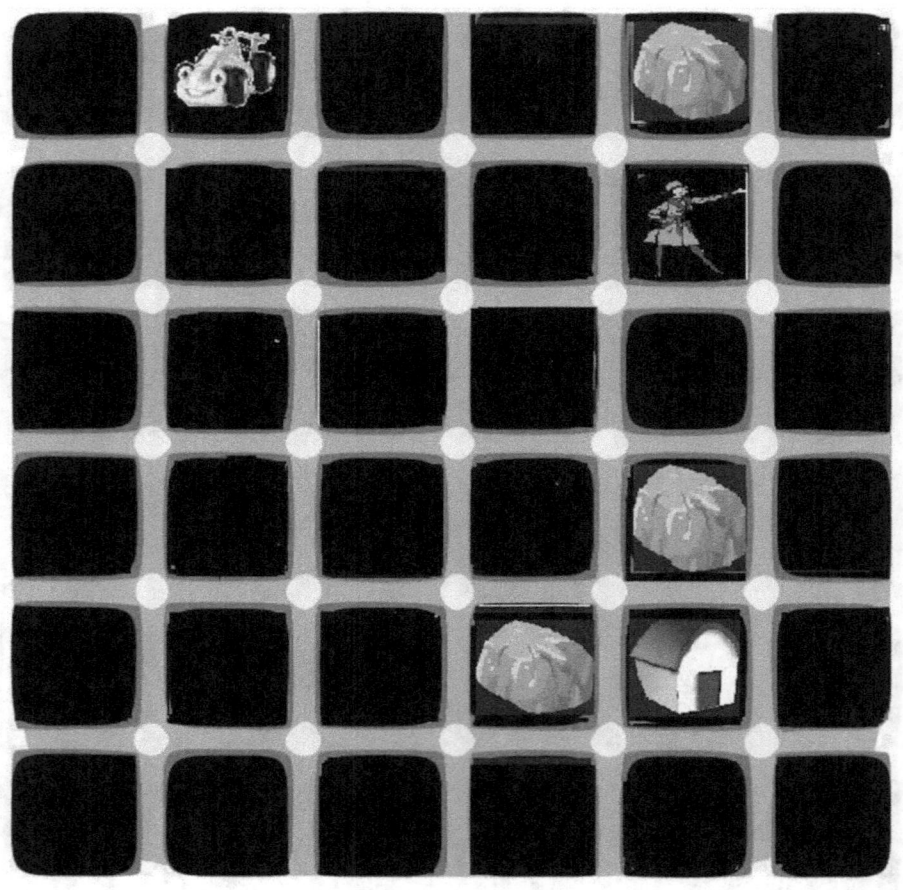

Challenge 26

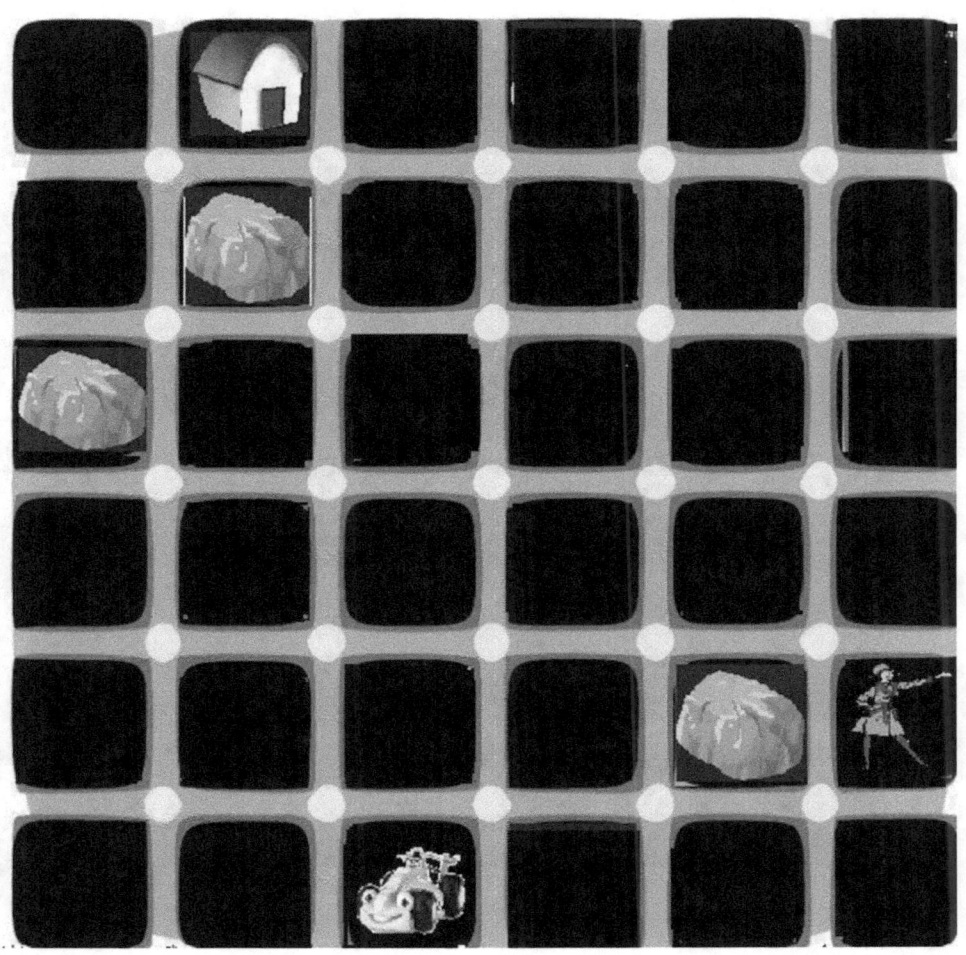

Challenge 27

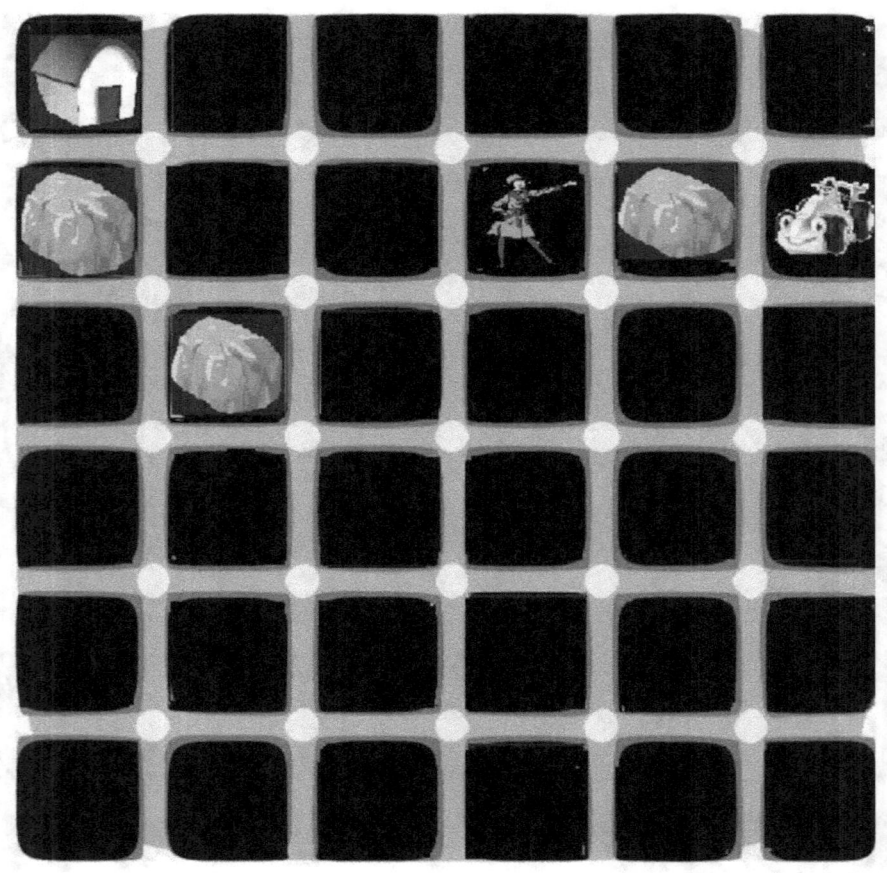

Challenge 28

Challenge 29

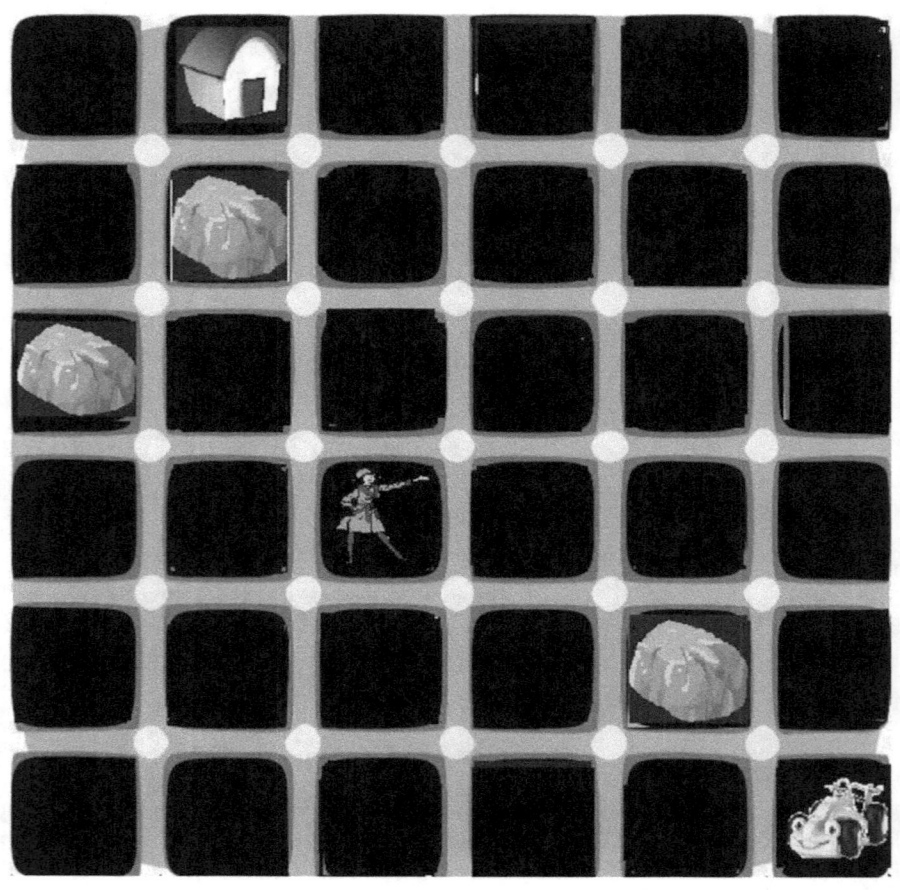

Challenge 30

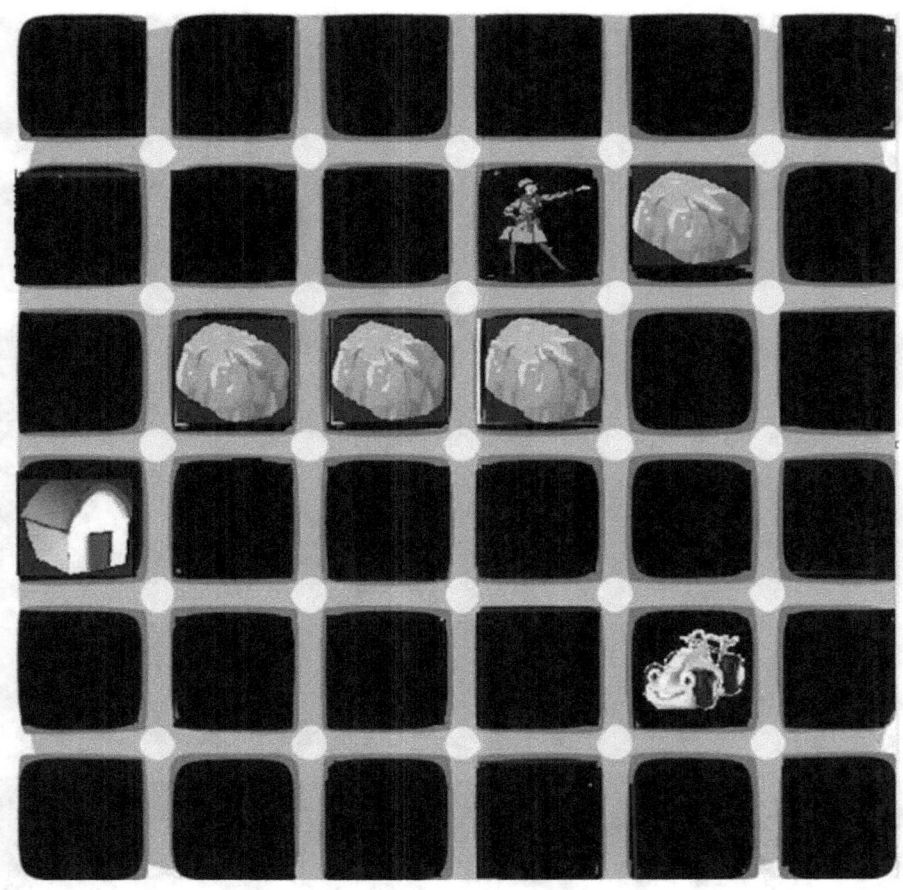

Challenge 31

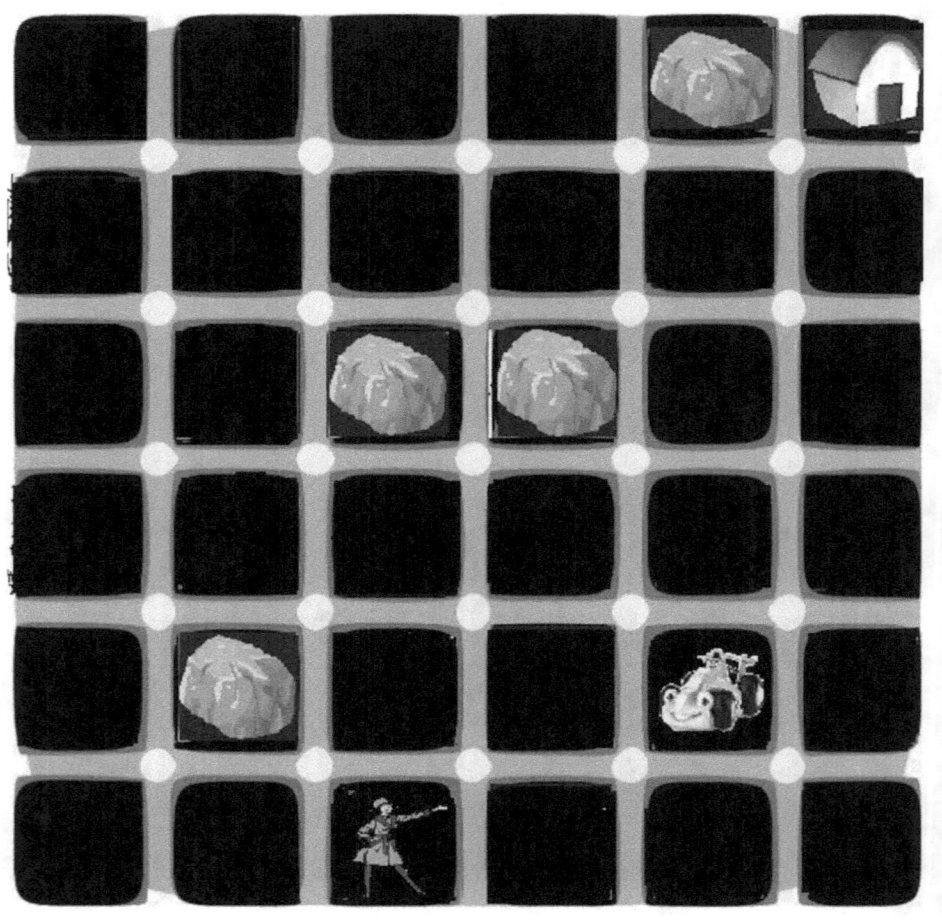

Challenge 32

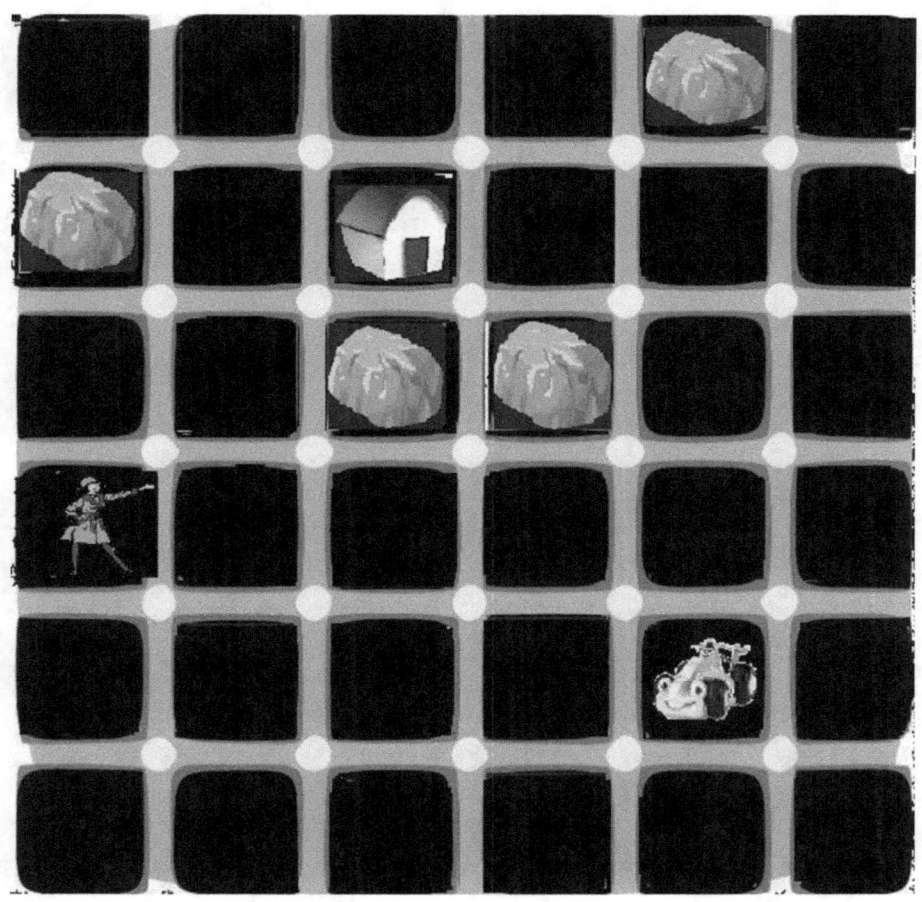

Challenge 33

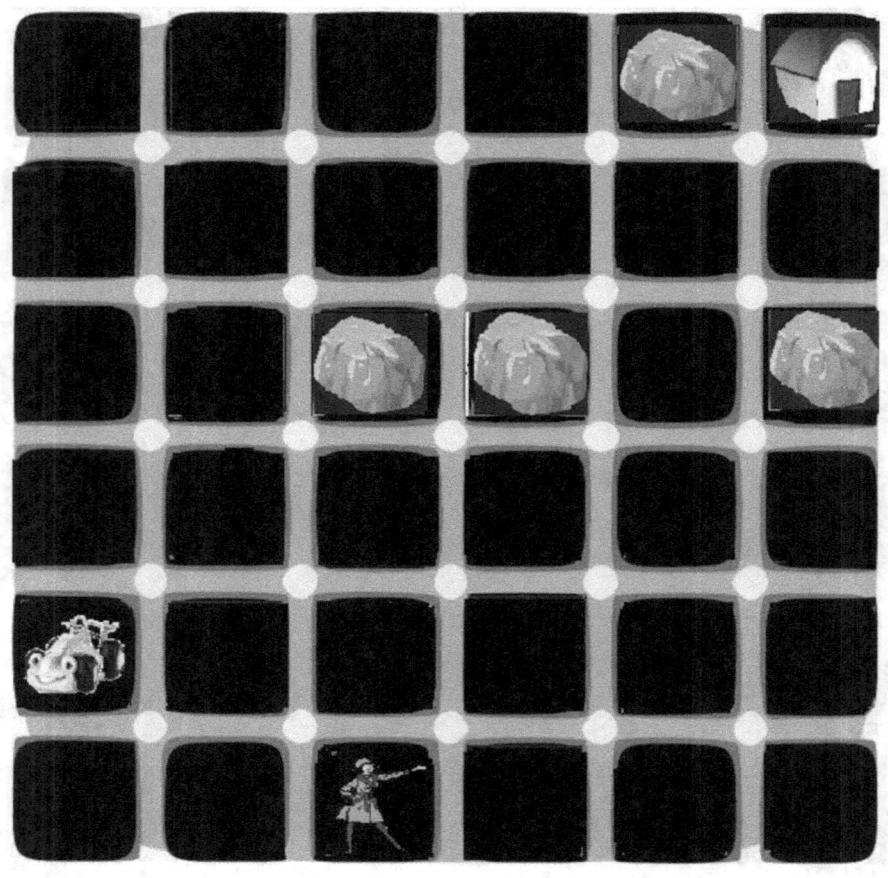

Challenge 34

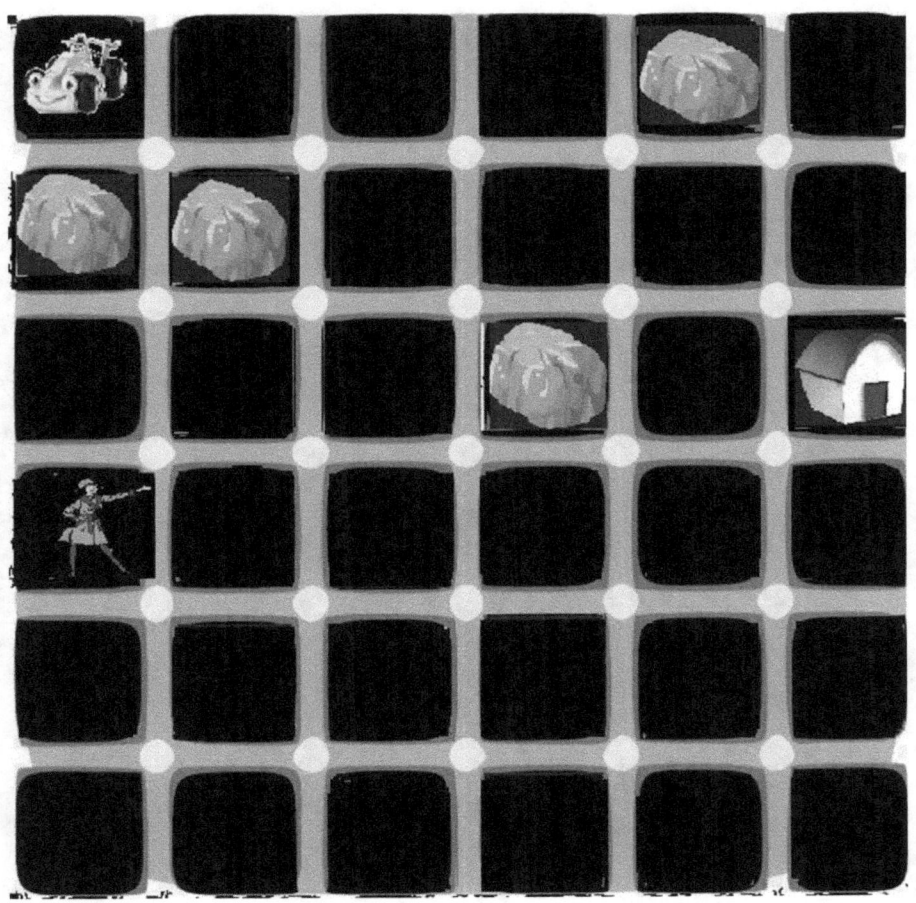

147

Challenge 35

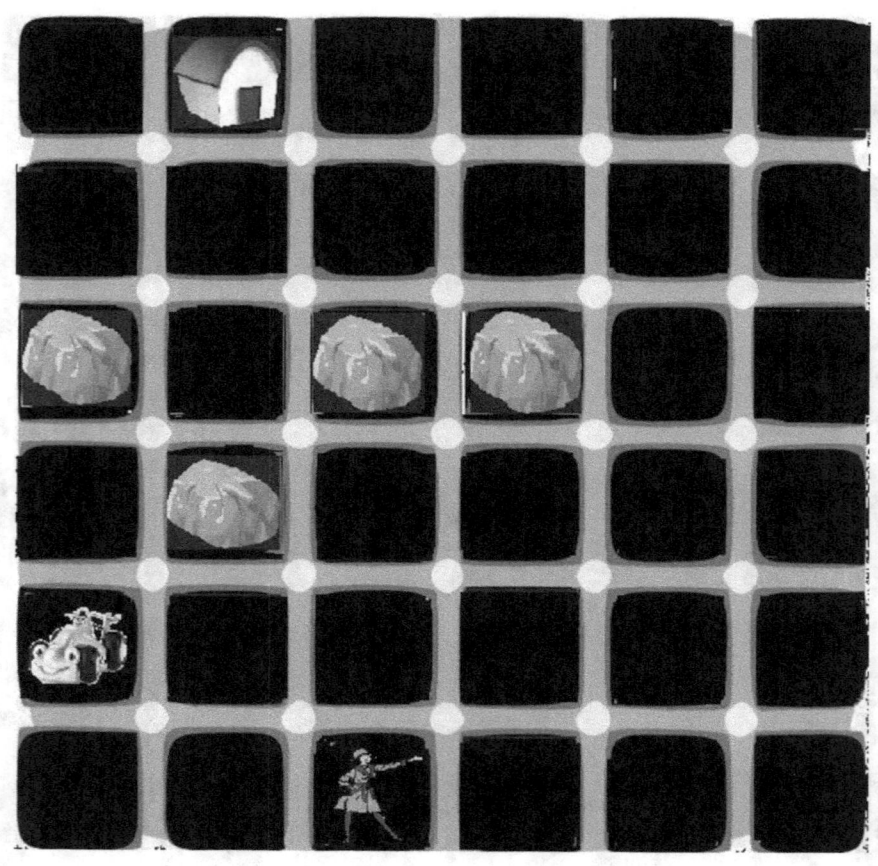

Number Key Challenge

Memorize the keys below and solve the problems that follow. Each digit corresponds to a picture, and this is used as a map for all the problems that follow.

Solve the below simple Math problems using the number key:

Challenge 36-40

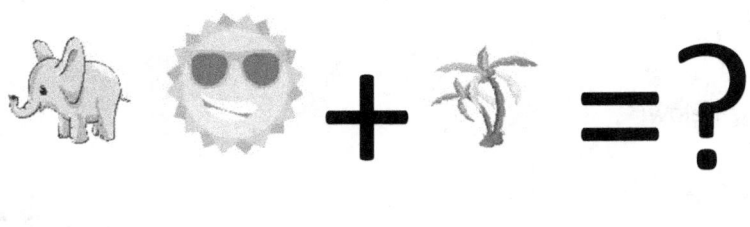

Challenge 41-45

Solve these below:

  X

Challenge 46-55

Find the squares of the following numbers:

Challenge 56-61

Find the remainder below. "%" is the remainder of the first number divided by the second.

For example, 5%2 is the remainder when 5 is divided by 2. So, 5%2 = 1.

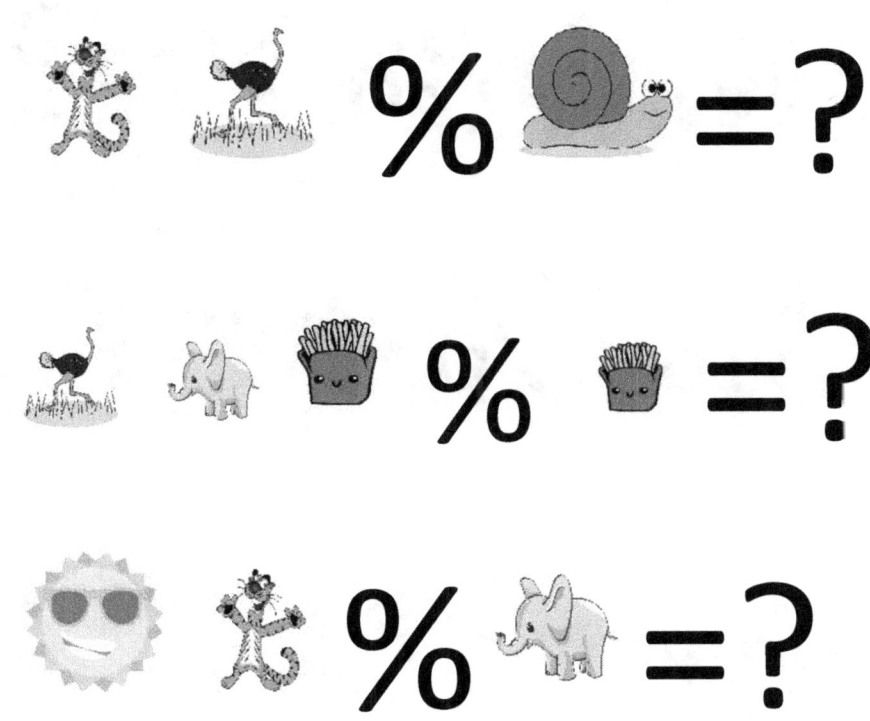

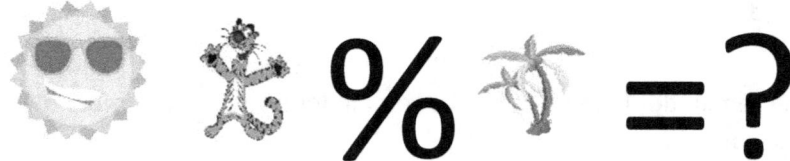

Challenge 62-67

Solve the problems below:

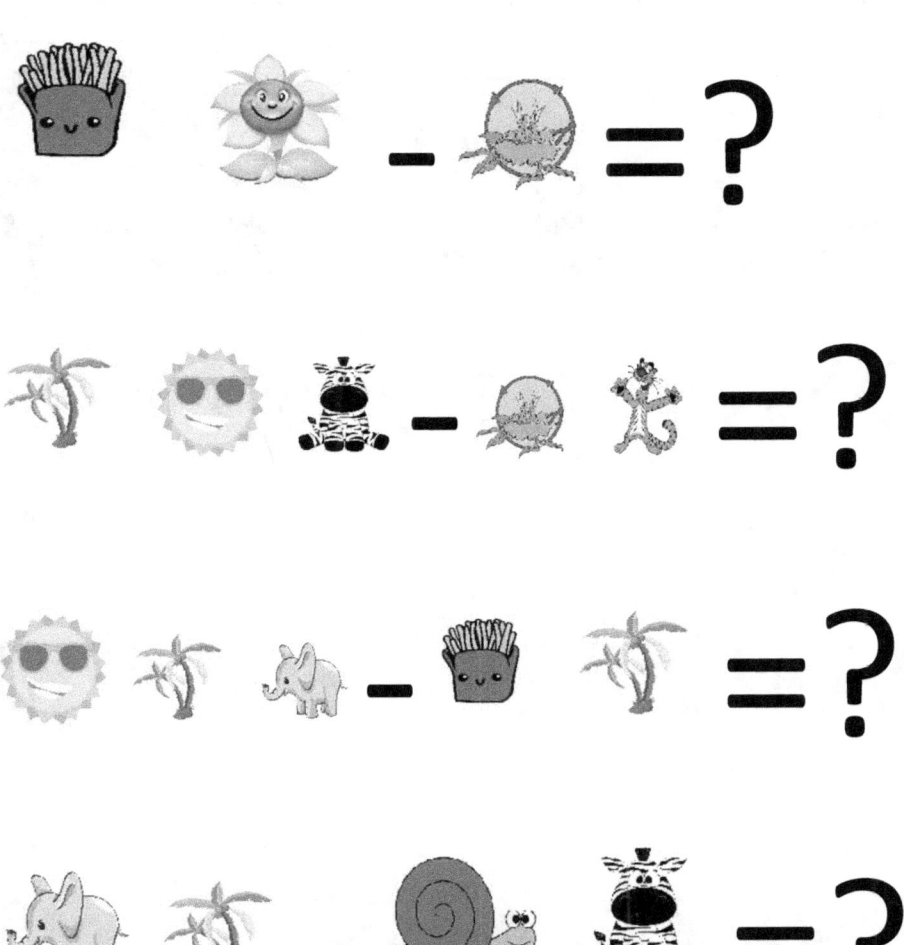

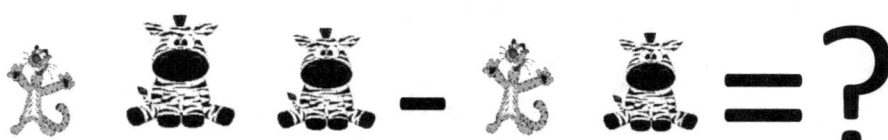

Challenge 68-72

Answer the following multiple-choice questions:

68. Which of the following is not a multiple of 11?

69. Which of the following is a palindrome?

70. Which of the following is a perfect square number?

71. Which of the following is the square root of 16?

72. What year was this book created?

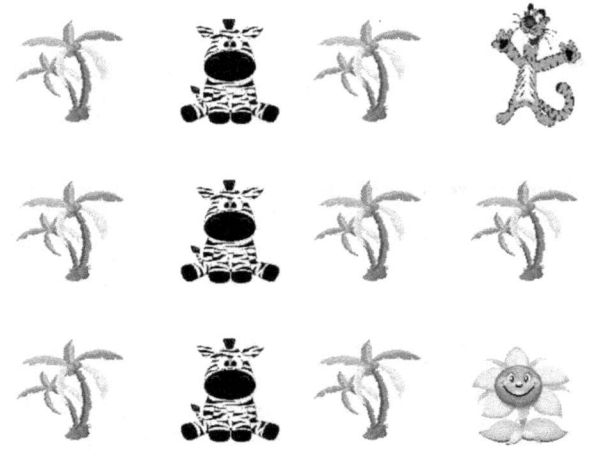

Loops

Imagine you have a toy train, and you want it to go around a toy track multiple times. A loop is just like that track. It assists you in doing something several times without having to repeat the same steps every time.

Let's say you want the train to go around the track five times. Instead of saying, "Okay, train, go around the track one...two...three times...four times...five times," you can use a loop. It's like a special command that tells the train to keep going around the track until it completes five rounds.

So, you can say, "Hey, train, go around the track five times!" The train starts moving and goes around the track once, twice, three times, four times, and finally, five times. The loop helps you avoid saying the same thing repeatedly.

Loops are like magic instructions that make things happen over and over again without getting tired or bored. They are helpful when you have a lot of things to do, and you want the computer or toy to do them automatically.

Let's say you want to print a pattern on the computer screen 1000 times. Without loops, you would have to copy and paste the print instructions again and again till you reach a count of 1000. However, with loops, you can just make a print instruction and put it inside a loop of 1000.

That's what loops are, like a special trick to make things repeat without you having to say or do the same thing again and again.

Example

Now, let's look at a loop with the same example where we print the character "&" 1000 times.

Now, if we didn't use a loop, we would do the following:

print "&"

print "&"

print "&"

print "&"

print "&"

......

Keep going until we do this a 1000 times.

Now, with a loop, we get:

Start of Loop from 1 to 1000:

print "&"

End Loop:

Elements of a Loop

There are 5 main elements of a loop:

A counter

A counter is an item that counts how many times the loop is executed.

Start Point

The start point is the number that the counter starts with.

End Point

The end point is the number that the counter ends with, after which the loop is no longer executed.

Increment

The increment is how much the counter is incremented every time the loop runs.

Loop Contents

The loop contents are the items inside the loop that are executed every time the loop is run.

Now, let's have a look at above elements given the same example again.

Start Loop: ctr = 1 to 1000:

print "&"

ctr = ctr +1;

End Loop:

In the above loop, ctr is the counter. The loop start point is 1 and end point is 1000. The increment is 1.

Here is how the loop works in pictorial form:

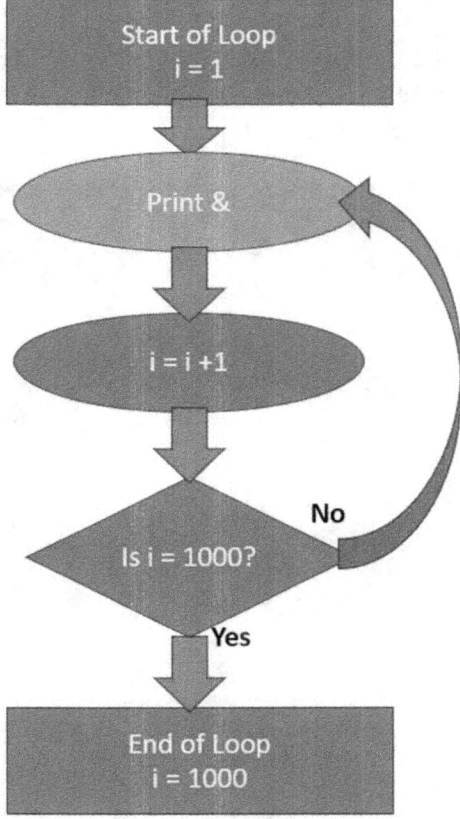

Now, let's have a look at a few simple examples of loops that we can use.

Challenge 73

Complete a loop that prints out all numbers from 1 to 20. The flowchart below that is supposed to solve the problem has missing blanks. Complete the missing blanks in the visual below:

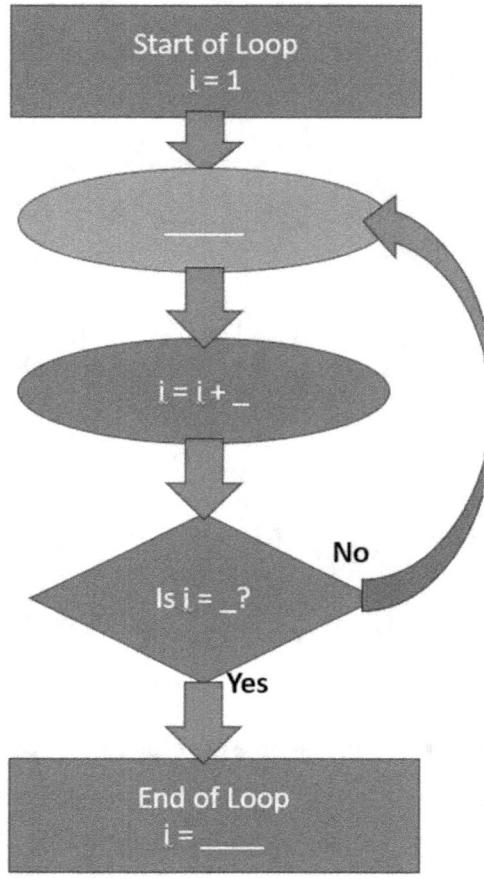

Challenge 74

Complete a loop that prints out all the odd numbers from 1 to 30. Complete the missing blanks in the visual below that aims to solve the problem.

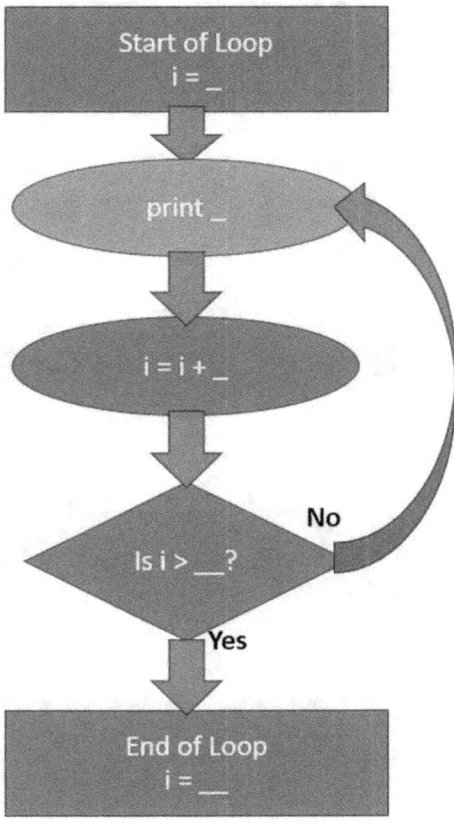

Challenge 75

We need a car to loop around a track 10 times. Part of the car loop is shown in the visual below. Complete the missing blanks in the visual below:

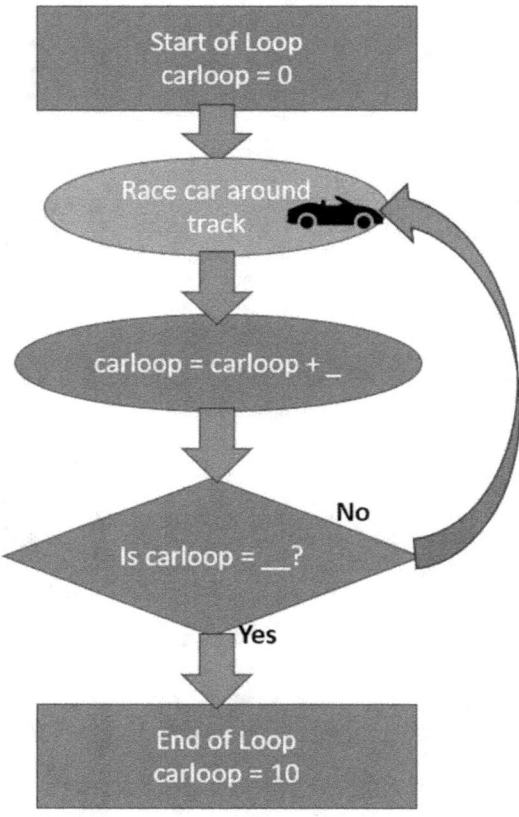

Challenge 76

You oversee a Mars space program. You have a spacecraft that needs to orbit around Mars, but it needs to only do so 5 times. Part of the orbit program is shown in the visual below. Complete the missing blanks in the visual below:

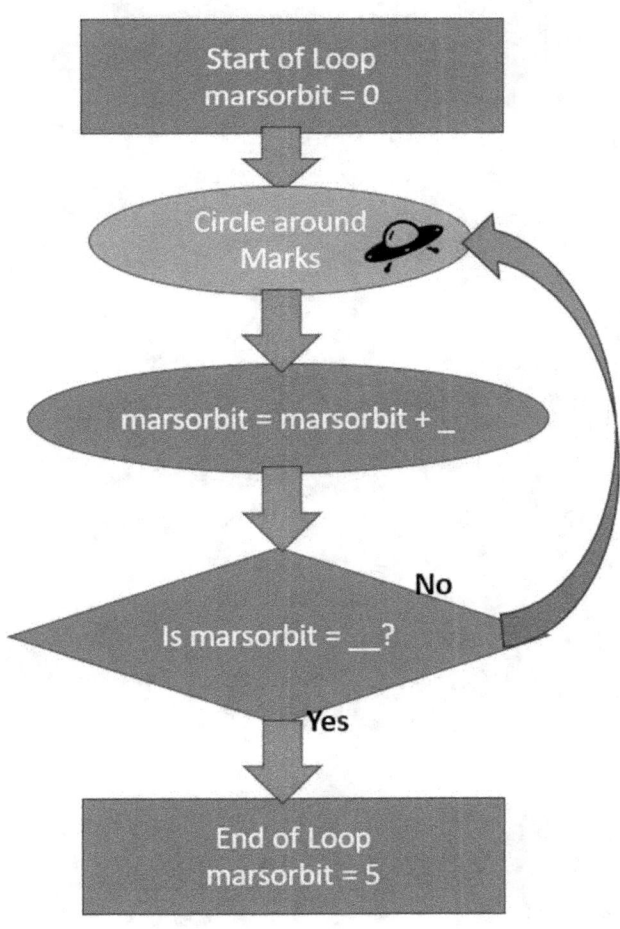

Challenge 77

You are tasked with printing a pattern on the screen 100 times. The pattern has the three characters "(*)". It needs to be printed on a separate line each time. Part of the program is filled out below. Fill out the rest to complete it.

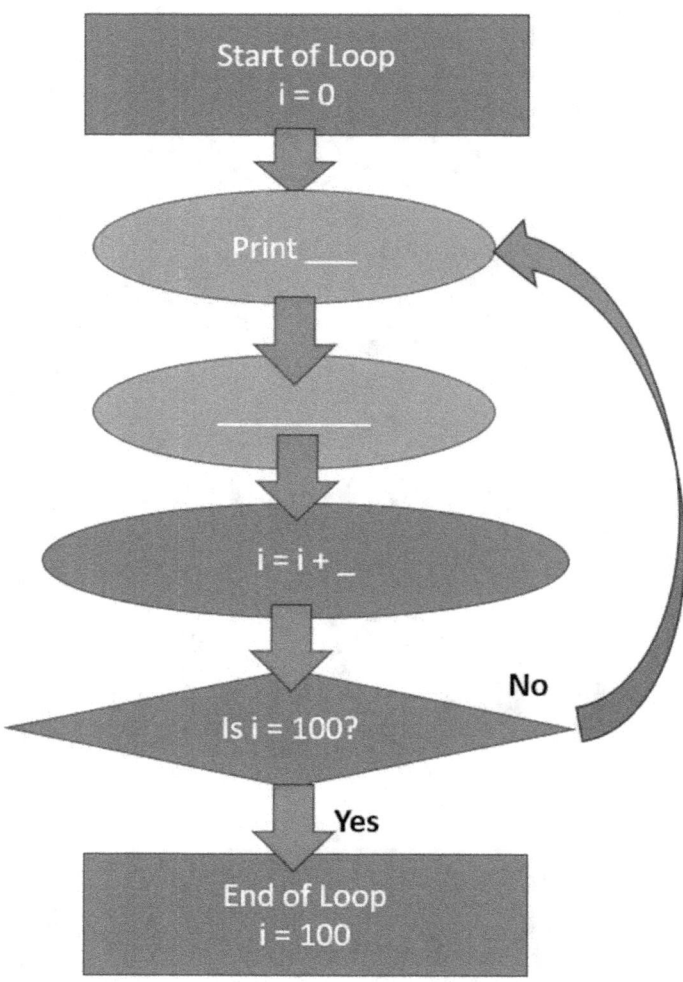

Nested Loops

Now, let's have a look at nested loops. Nested loops are basically loops within a loop. To explain this, let's say you have a box of toy spacecraft. There's supposed to be 10 spacecrafts. Each spacecraft is made of 5 parts. So, you check each toy spacecraft for its 5 parts, then move on to the next one. You do this 10 times till you run out of spacecrafts.

So, what you just completed is a nested loop. The outer loop is the counting of the spacecraft 10 times. Within each loop of counting spacecraft, there is a loop where you check the 5 parts of each spacecraft. This is the inner loop.

The nested loop example is shown below. The green loop is the inner loop where we check the 5 parts of the spacecraft. The blue loop is the outer loop where we count the number of spacecrafts.

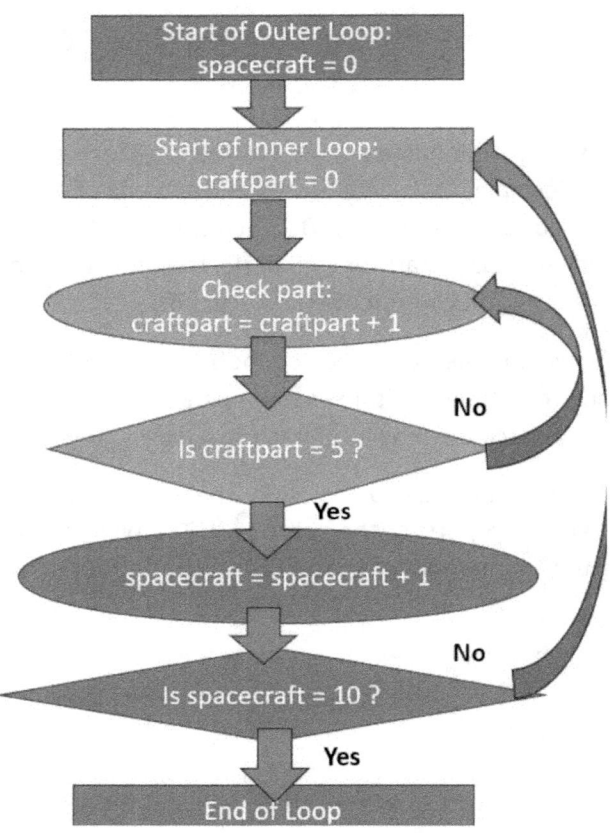

Challenge 78

A car needs to go around a racetrack 100 times. The way you can tell if it completes a lap is when the car completes 3 bends, as each lap has 3 bends in it. Use a nested loop to keep track of the number of times the car goes around the track, till it reaches 100 laps.

The below diagram helps achieve this objective, but it is missing a few blanks. Fill in the blanks to complete this task.

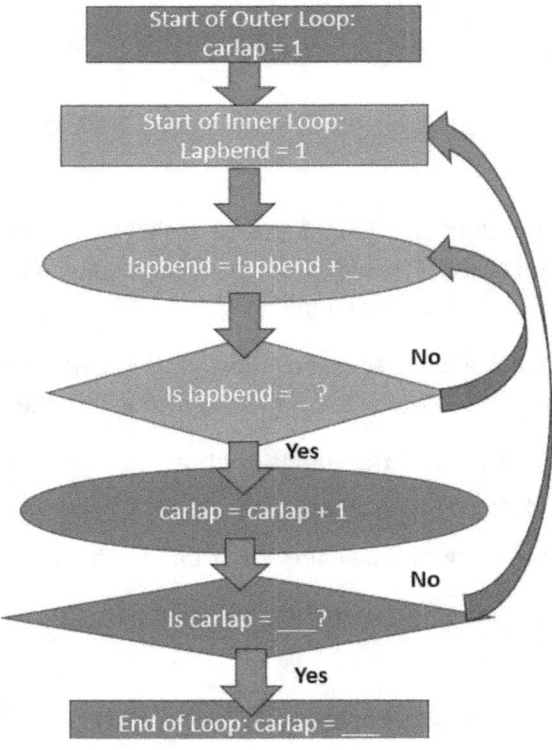

Challenge 79

Print the sum of numbers from 1 to a certain number. For example, if the number is two, print the number 3 (1+2). Do this for the first 10 numbers and move onto the next line each time. The output is below:

1

3 (1+2)

6 (1+2+3)

10 (1+2+3+4)

....

...

Below is the flowchart that completes the task. The outer loop has a counter called i that starts at 1. The outer loop also has a sum counter called sum which counts the values of the inner loop.

The inner loop has a counter called j. Each time j is increased, it is added to the value of the sum. When j is equal to the value of i, the inner loop ends.

i is incremented by 1, and the sum is printed onto the screen and the cursor moves to the next line.

Then we start the inner loop again with j=1 and sum=0.

The outer loop ends when i reaches a value of 10.

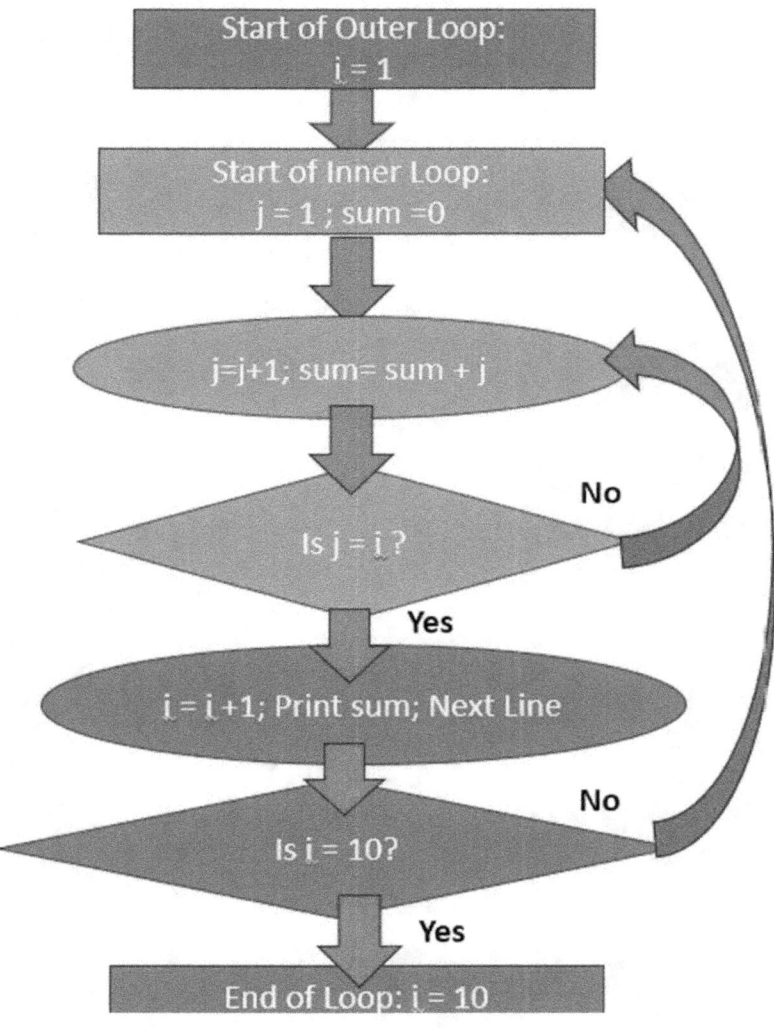

The challenge for this problem is to complete the output of the program above. We'll fill out the first three numbers, but you need to fill out the 7 numbers that follow:

Output:

1
3
6
—
—
—
—
—
—
—

Challenge 80

Print the numbers from 1 to 10 in a sequential fashion. For example, the first line is 1. The second line is 1 2. The third line is 1 2 3. This is best achieved using nested loops. The output is below:

1

1 2

1 2 3

1 2 3 4

1 2 3 4 5

1 2 3 4 5 6

1 2 3 4 5 6 7

1 2 3 4 5 6 7 8

1 2 3 4 5 6 7 8 9

1 2 3 4 5 6 7 8 9 10

Below is the flowchart that will make it happen. Fill in the missing blanks to complete this task:

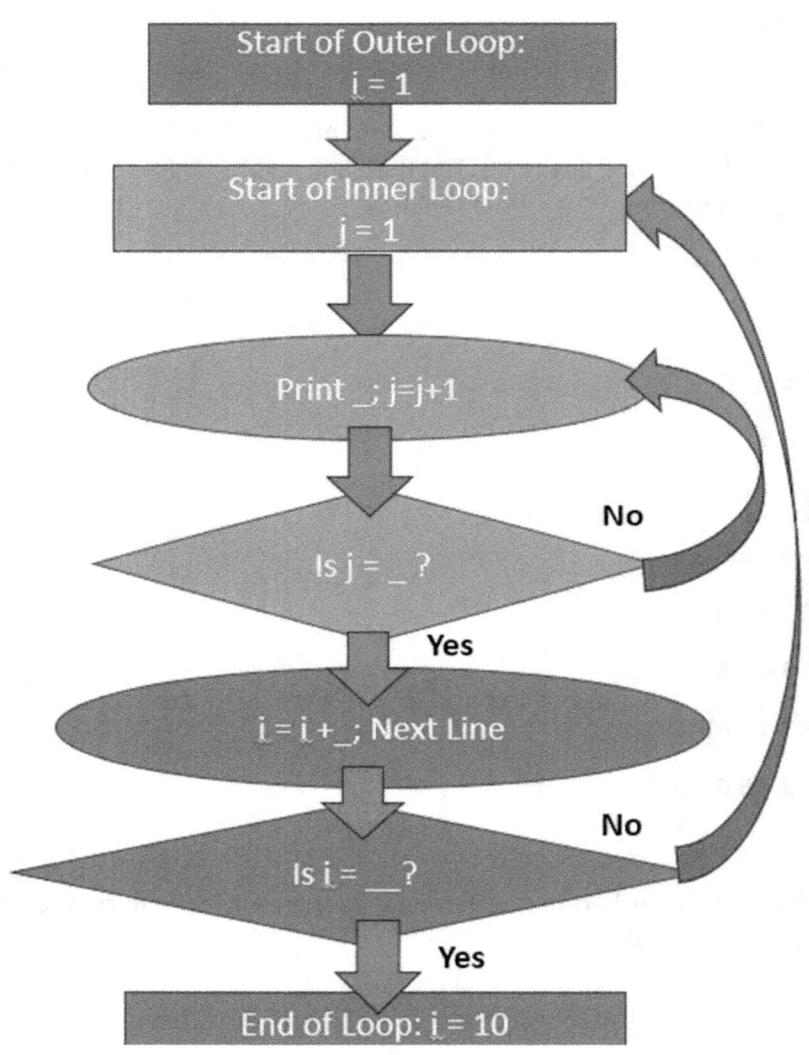

Challenge 81

Print this output on the screen below:

*

**

***

****

*****

So, there's 5 lines of this pattern. Complete the chart below that will achieve this goal.

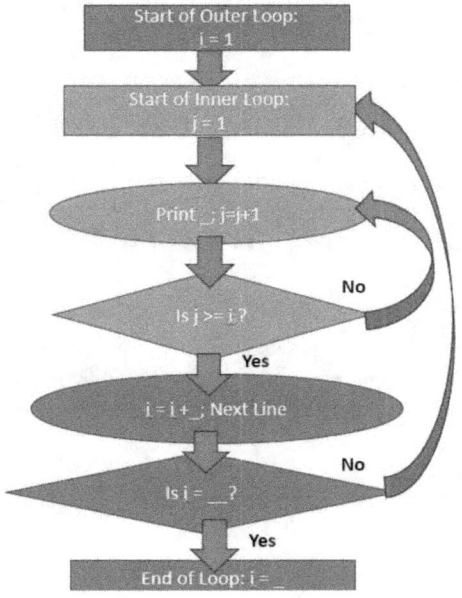

Conditional Loops

A conditional loop is a loop that is executed until a certain condition is met. For example, let's say that you want to orbit a spacecraft around Mars till it runs out of fuel. The loop orbits the spacecraft with a condition inside it. The condition is the check for fuel, and the spacecraft stops orbiting as soon as the fuel is below a certain level.

Here's an example of how the spacefuel conditional loop works.

The loop starts with a spacefuel value of 1000. The loop orbits mars and checks spacefuel. At the end of each orbit, spacefuel is checked to see if it is below 10. If it is below 10, the spacecraft leaves orbit to refuel.

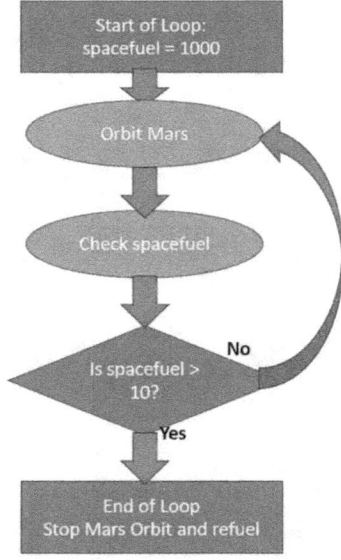

Challenge 82

Write a conditional loop for the toss of a coin. The loop keeps going until the number of heads is 5.

Below is a flowchart of the loop which is incomplete. Fill in the blanks to complete this.

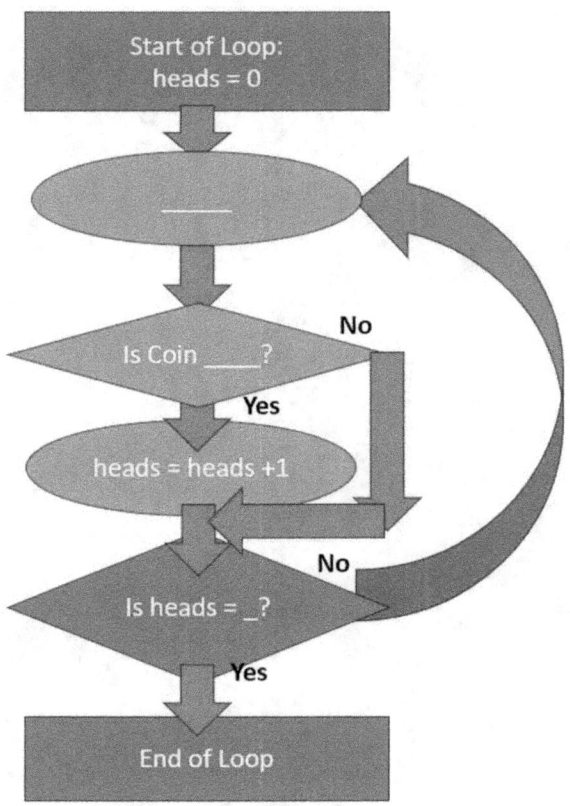

Challenge 83

Create a loop that checks a car's tires after each loop of a racetrack and stops when the car thread is below 2 mm due to wear, to go to a pitstop and change tires.

Below is the loop that is incomplete. Fill in the blanks to complete the loop.

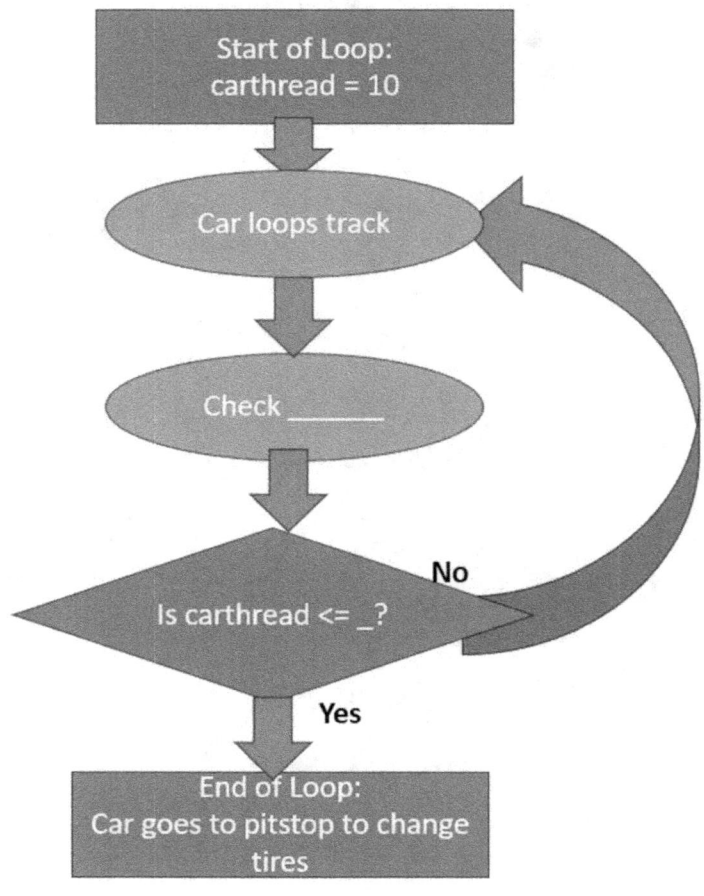

Challenge 84

You are working for a credit card company. You are tasked with designing a loop for each customer that alerts the customer when his credit balance is below $100. Design a loop that checks if a customer's account credit card balance is below $100 after each transaction. The loop stops when this happens.

The incomplete conditional loop is below. Fill in the blanks to complete.

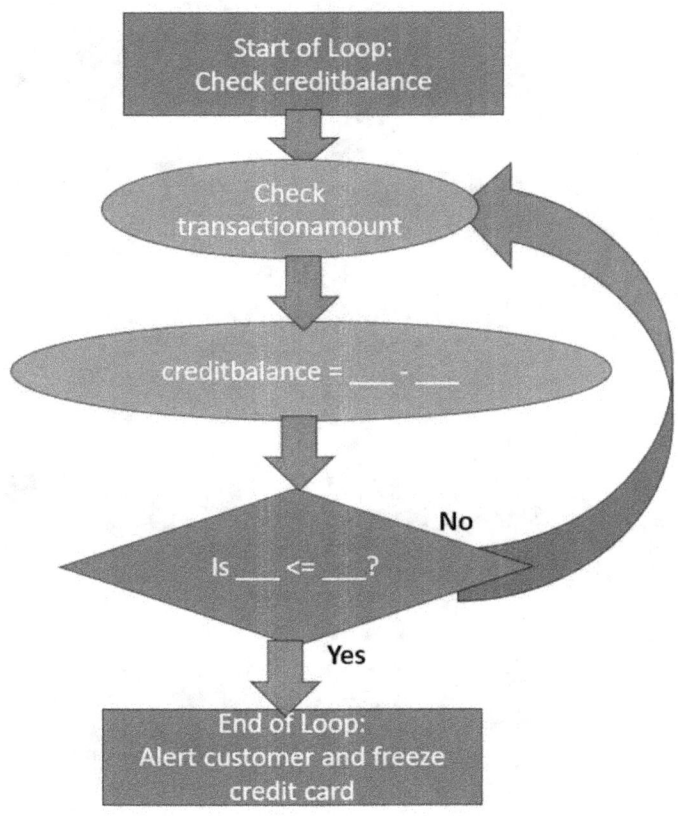

Programming Challenges

Find the outputs for the following loops:

Challenge 85

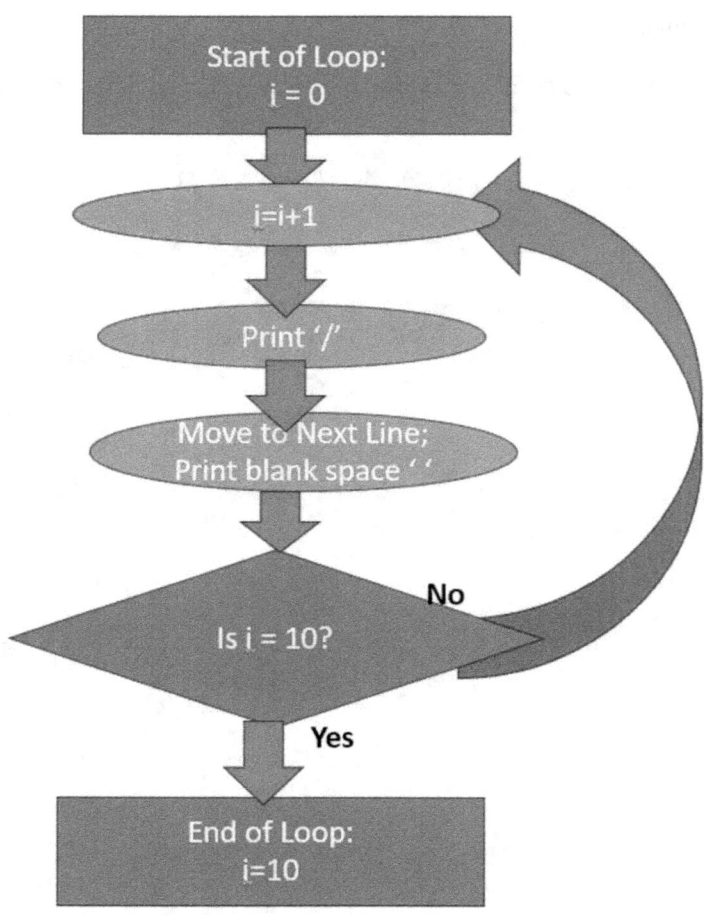

Challenge 86

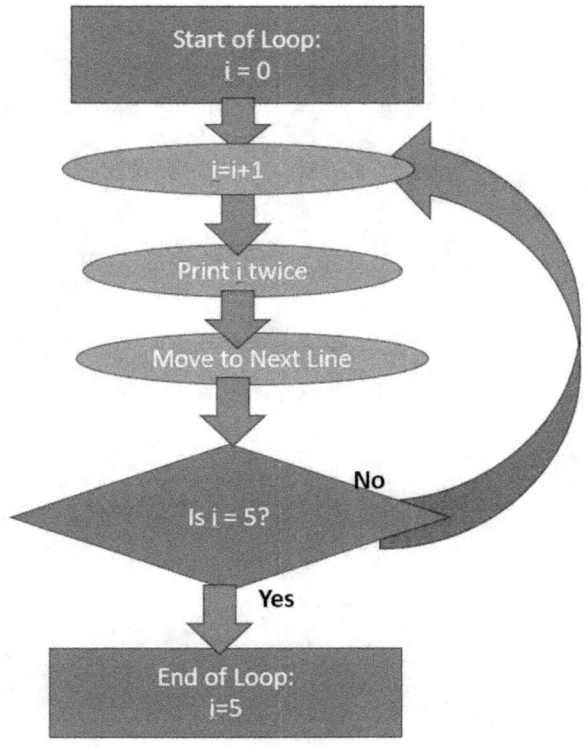

Challenge 87

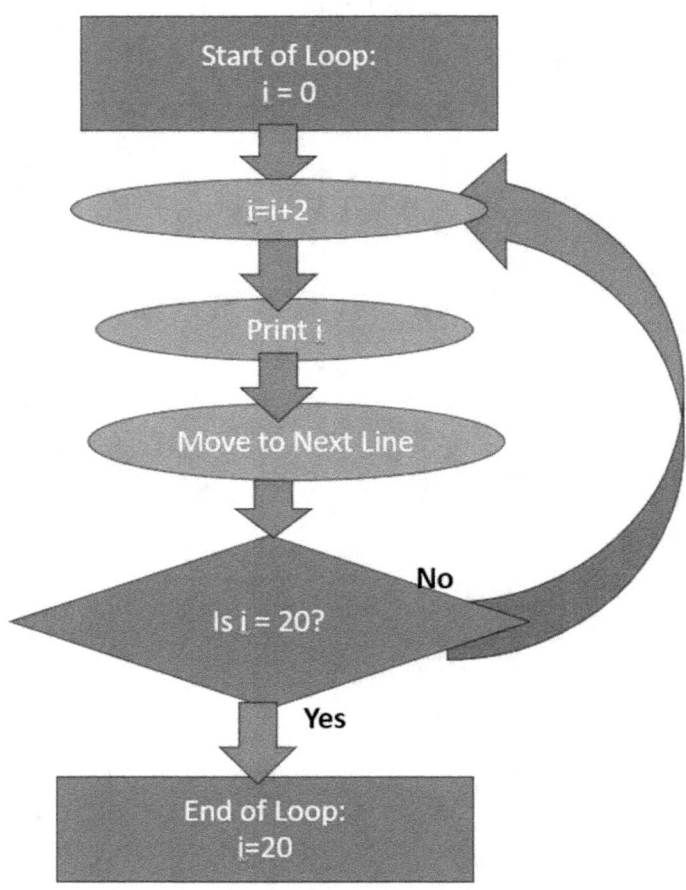

Challenge 88

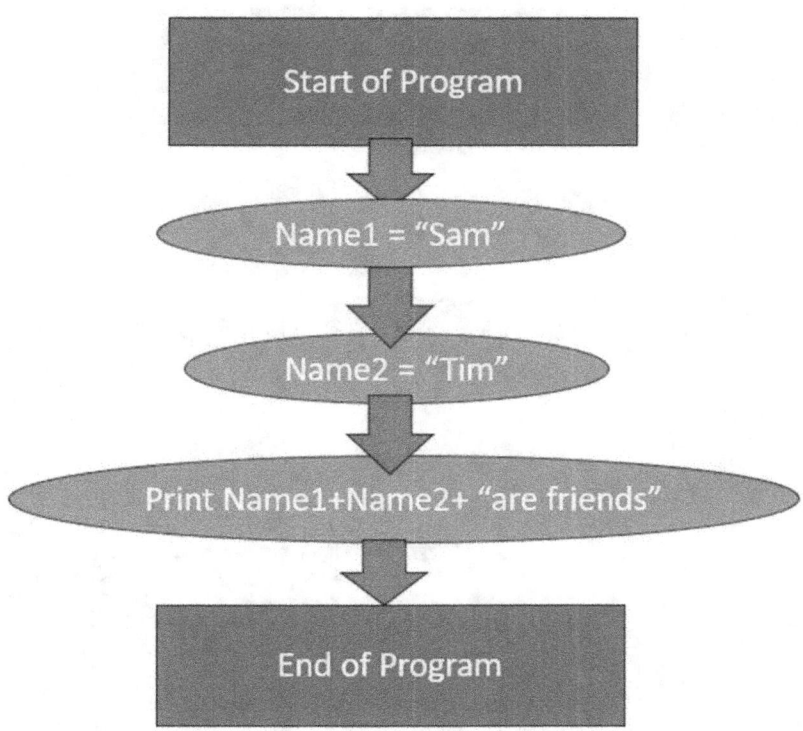

Challenge 89

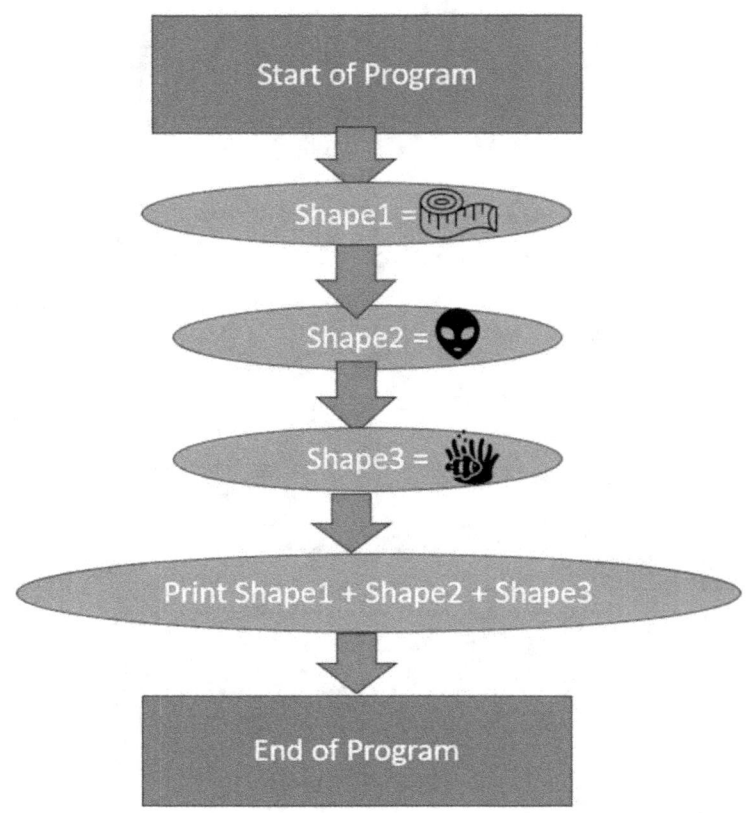

Challenge 90

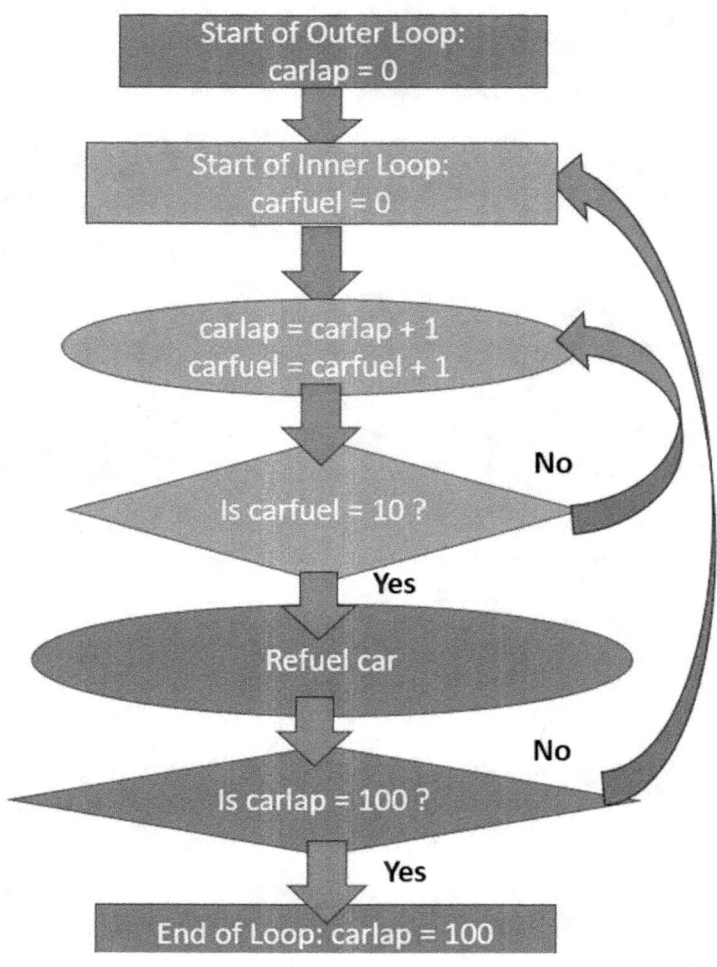

Challenge 91

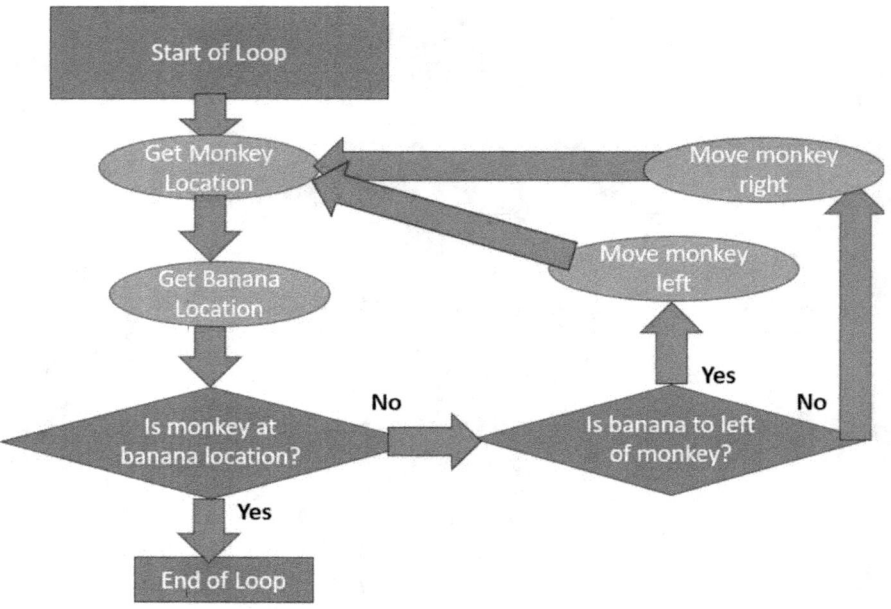

Challenge 92

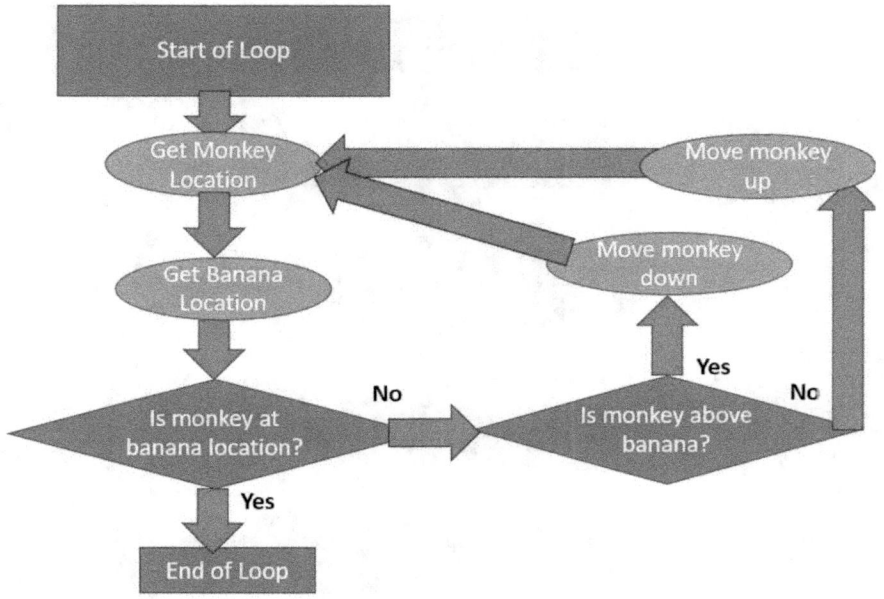

Challenge 93

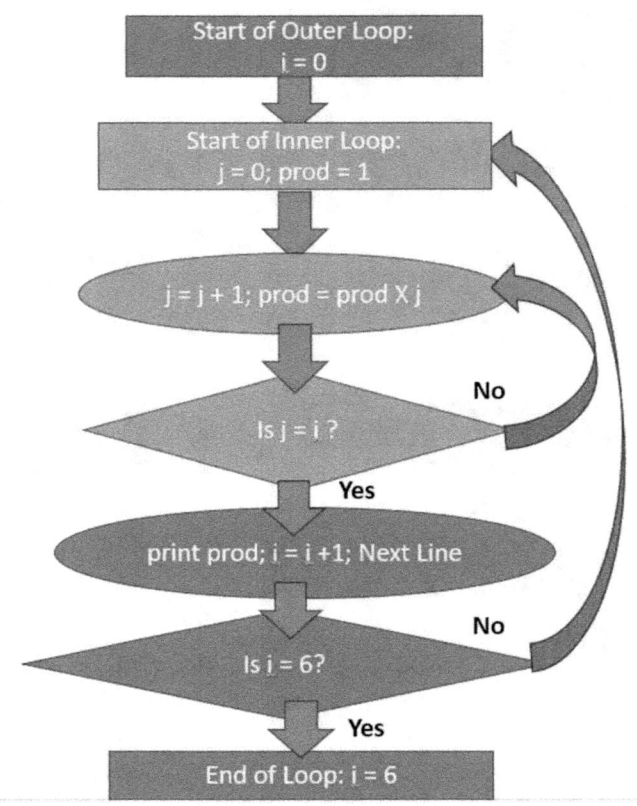

Answers

The Search for that Elusive Banana

Challenge 2

Challenge 3

Challenge 4

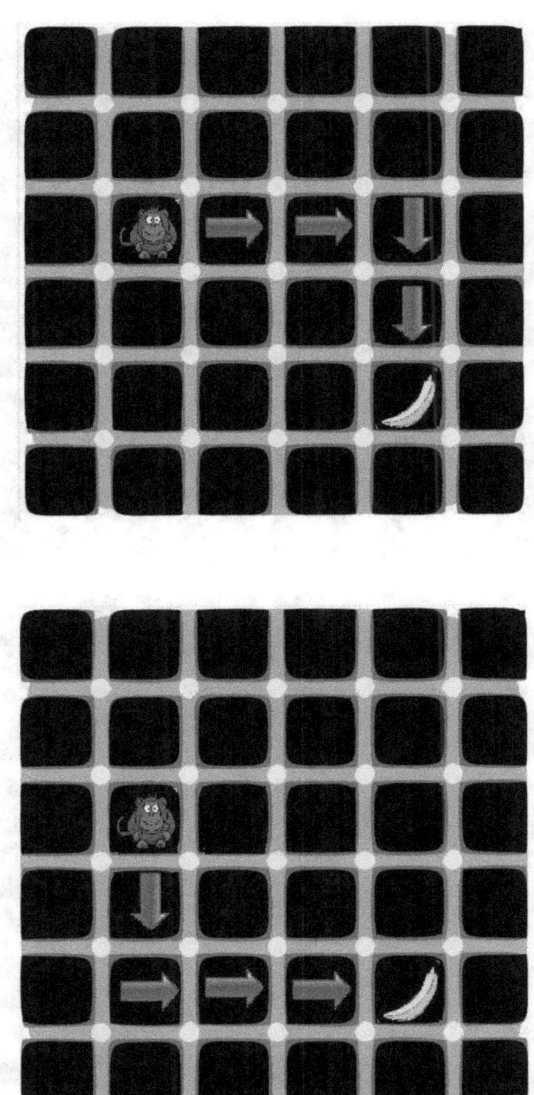

Challenge 5

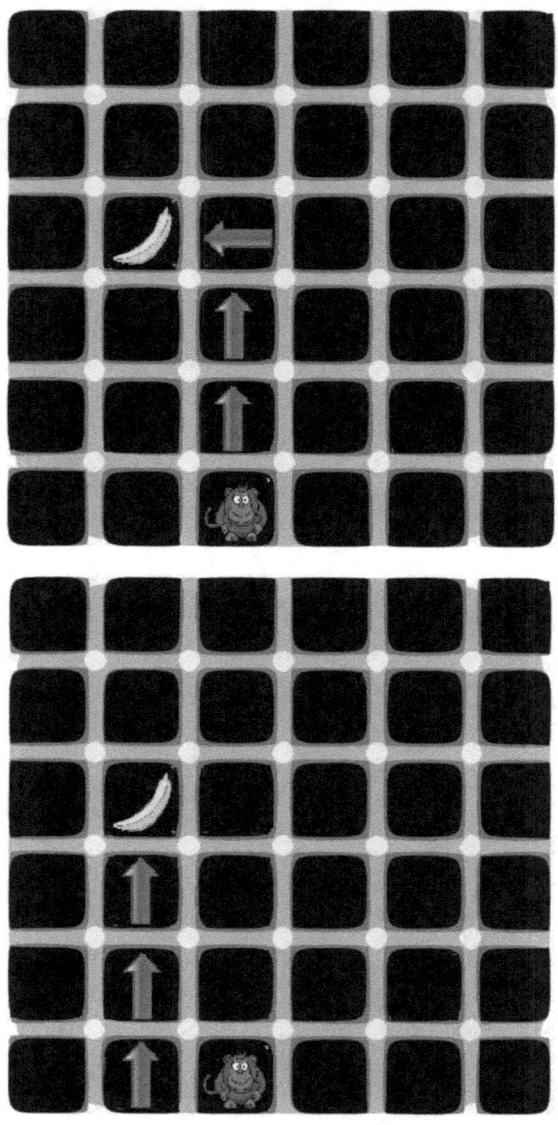

Challenge 6

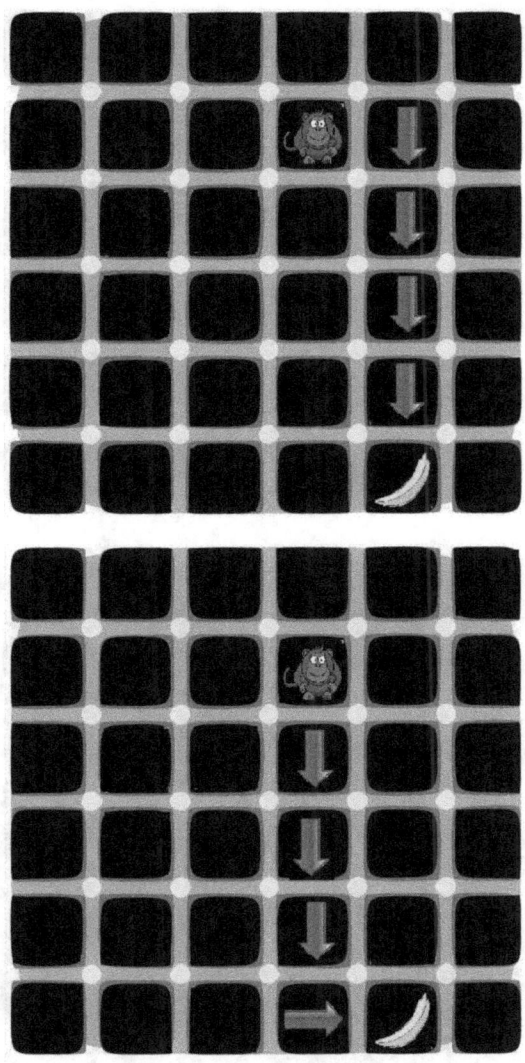

Challenge 7

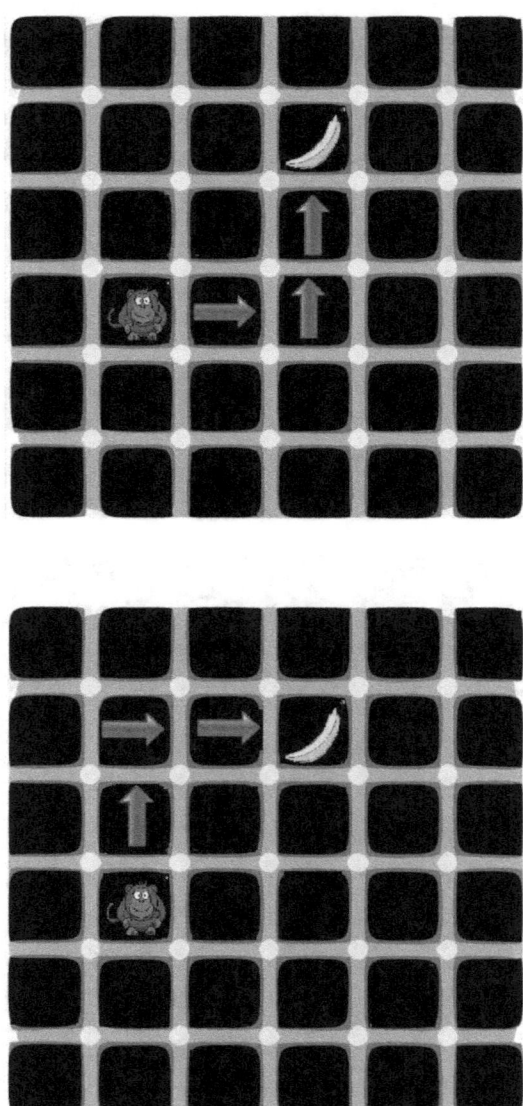

Challenge 8

Challenge 9

Challenge 10

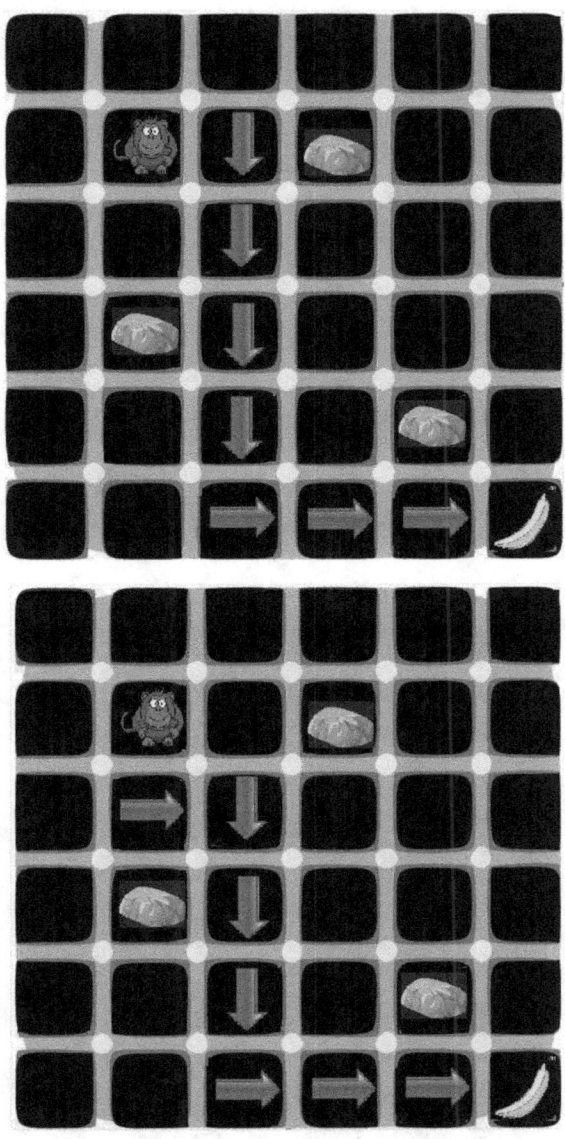

Challenge 11

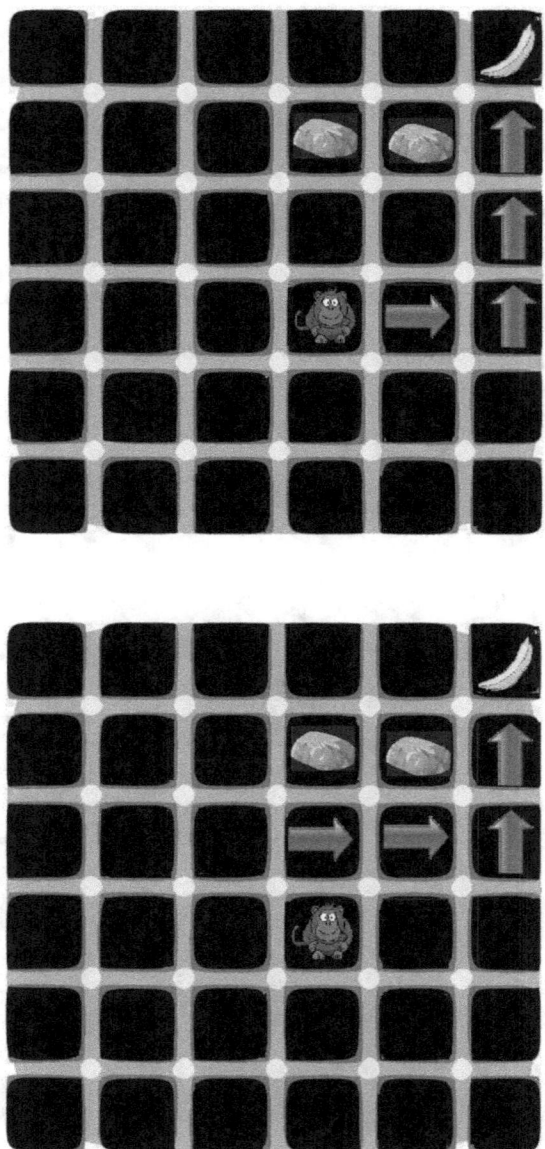

Challenge 12

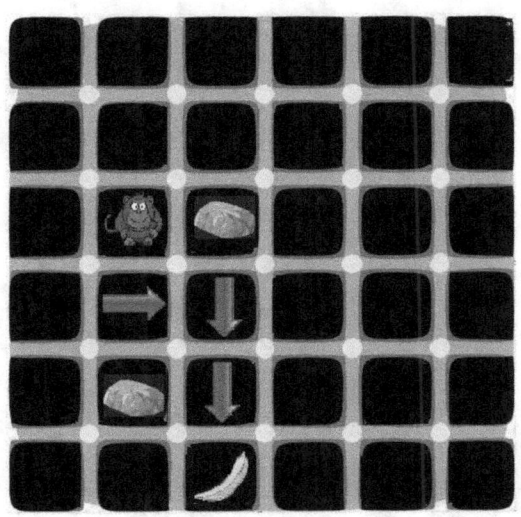

Challenge 13

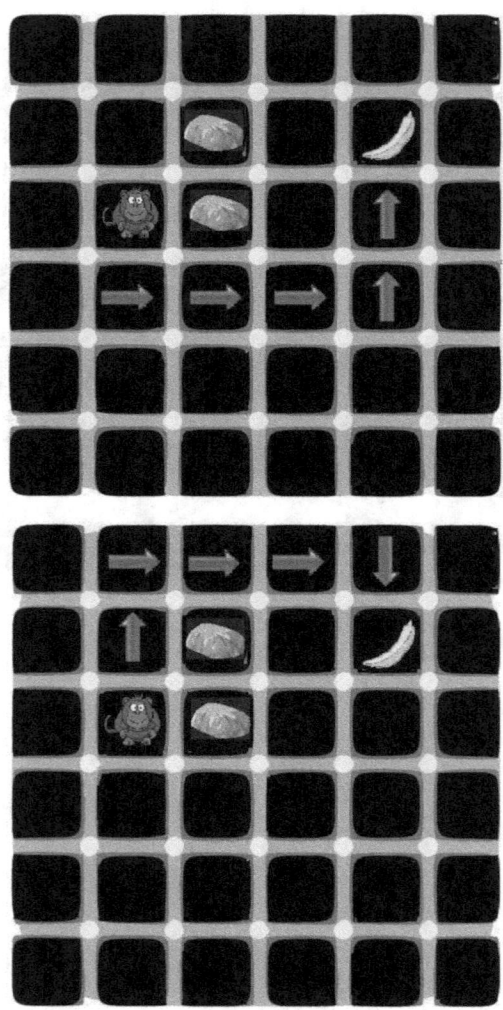

Challenge 14

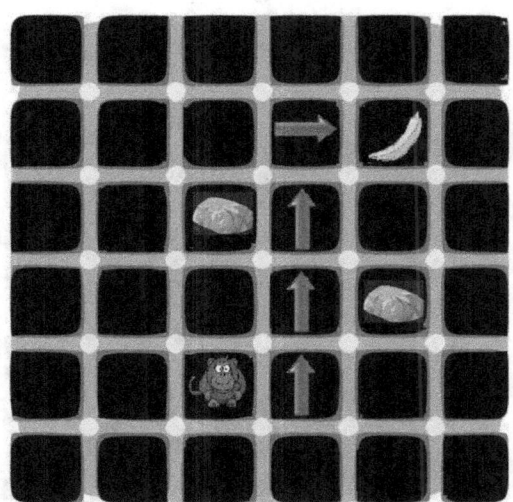

Challenge 15

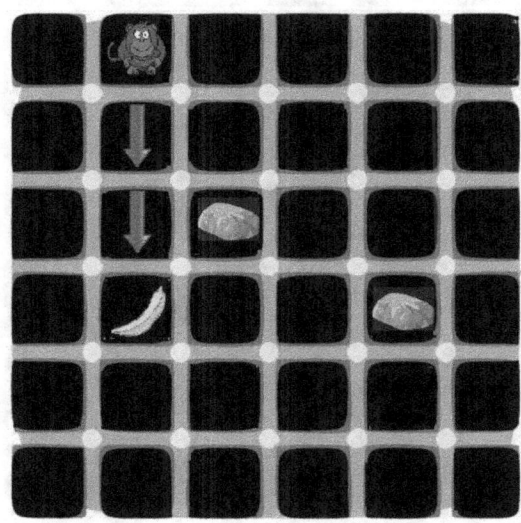

Taxi Please

Challenge 16

Challenge 17

Challenge 18

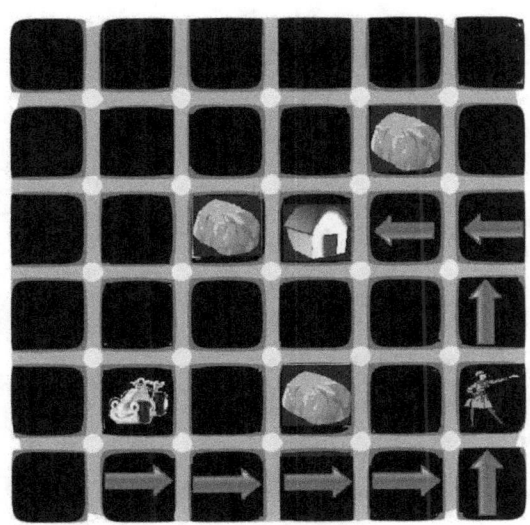

Challenge 19

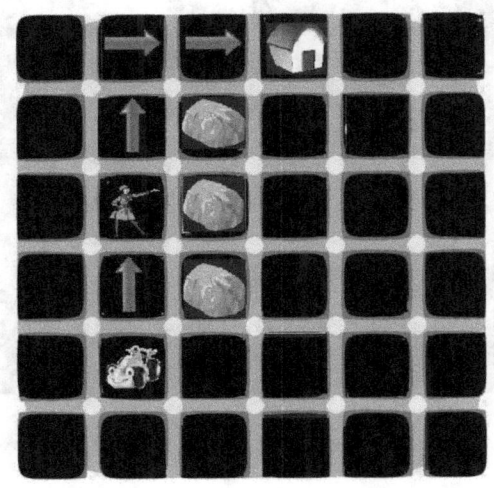

Challenge 20

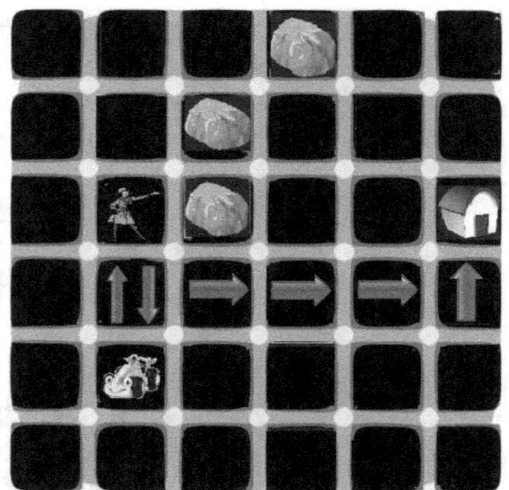

Challenge 21

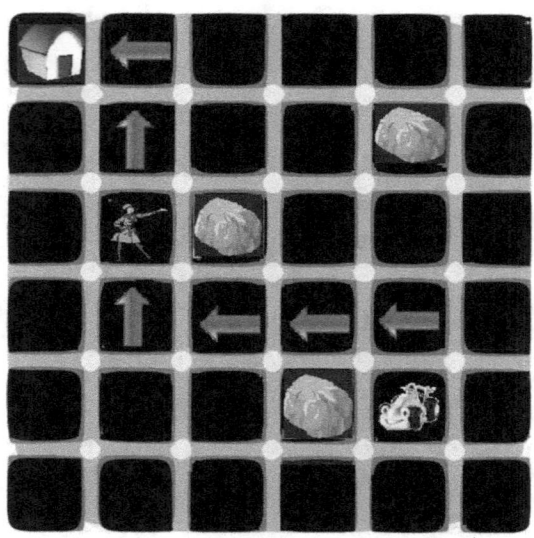

Challenge 22

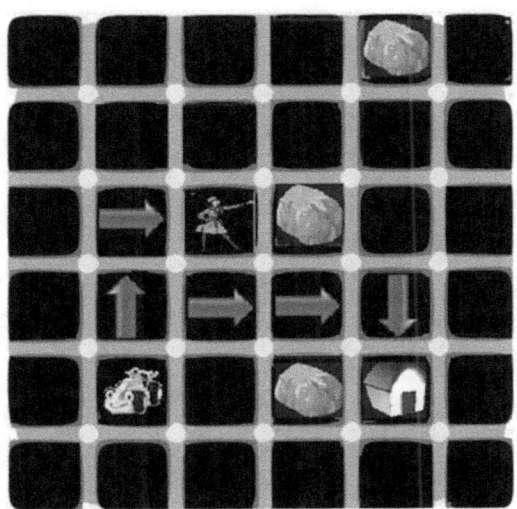

Challenge 23

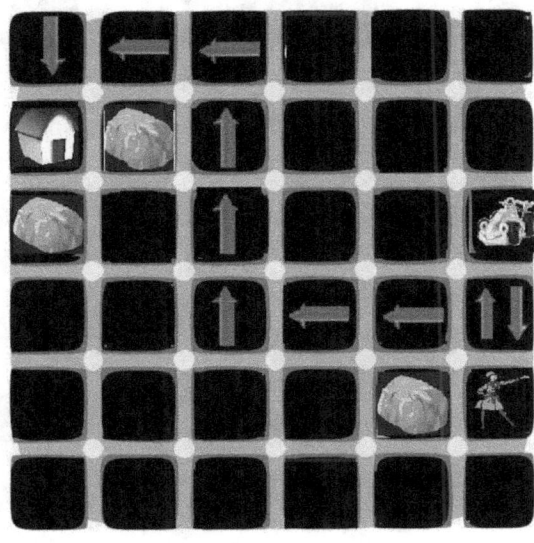

Challenge 24

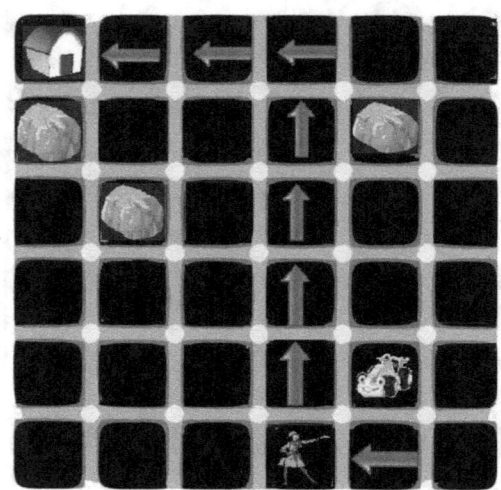

Challenge 25

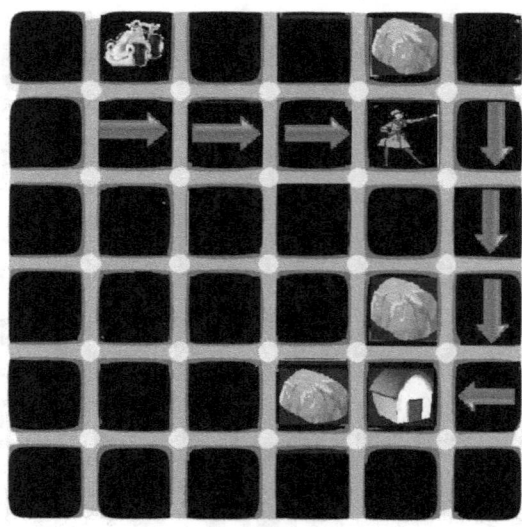

Challenge 26

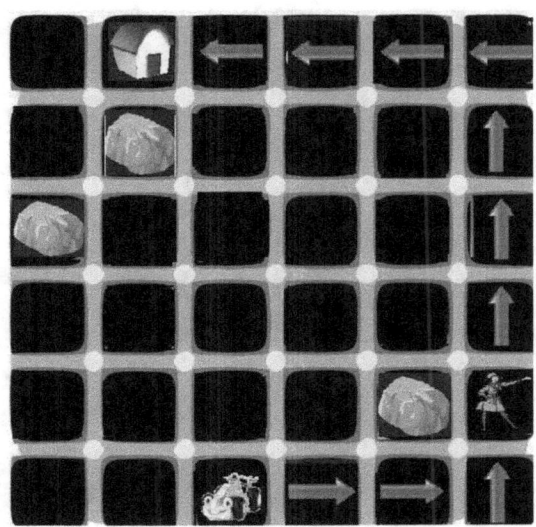

Challenge 27

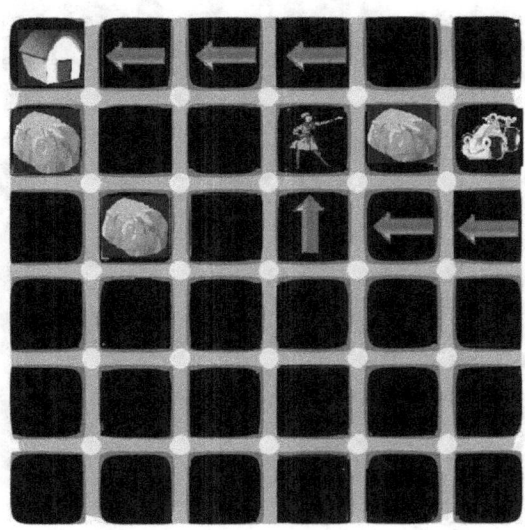

Challenge 28

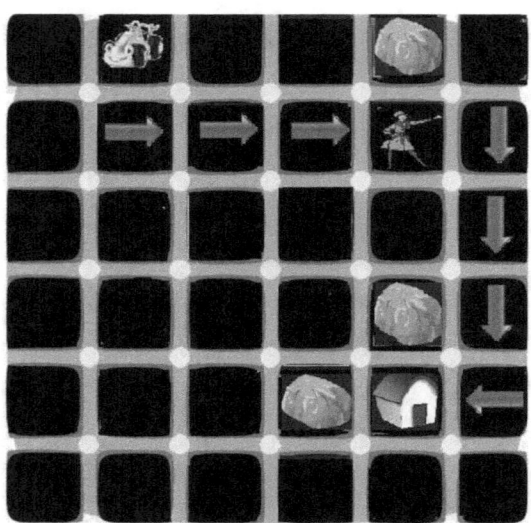

Challenge 29

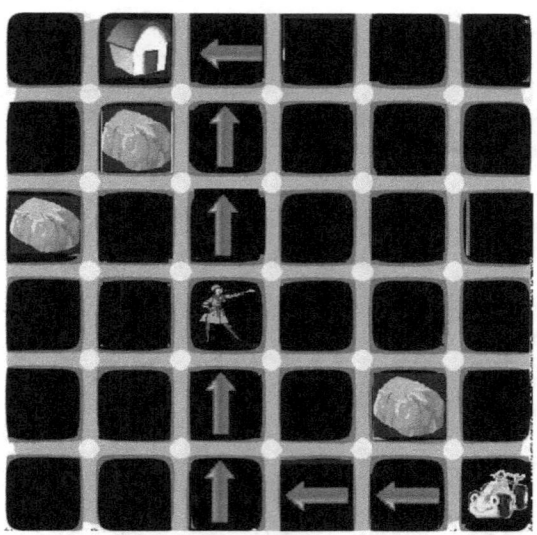

Challenge 30

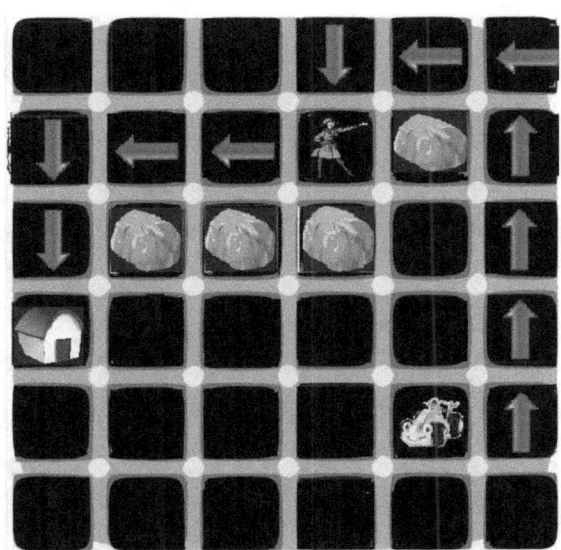

Challenge 31

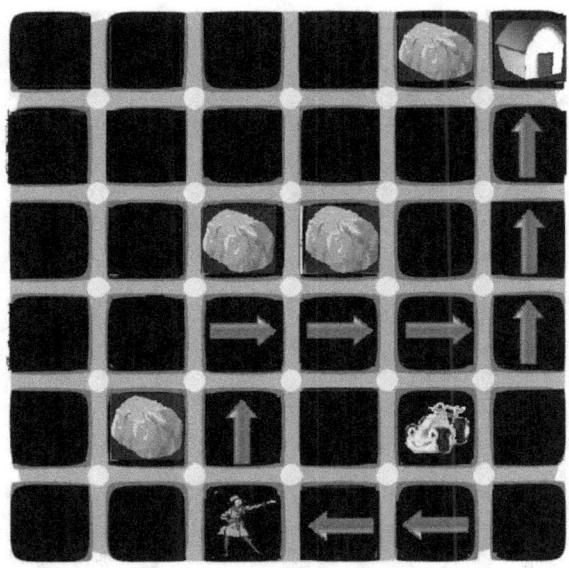

Challenge 32

Challenge 33

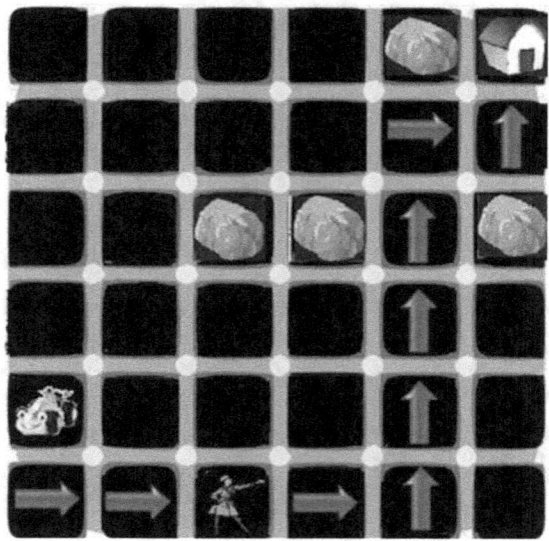

Challenge 34

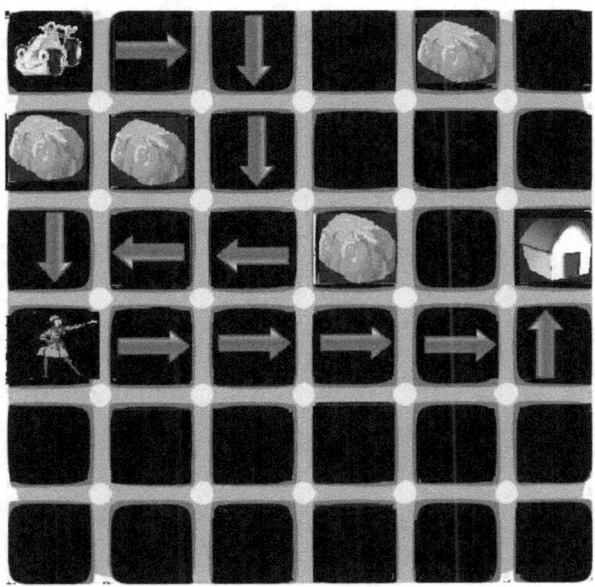

Challenge 35

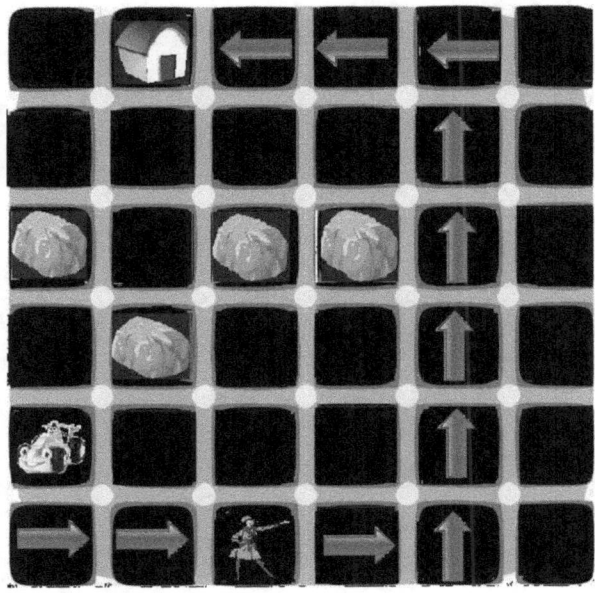

Number Key Challenge - Answers

36. 1 + 7 = 8
37. 37 X 2 = 74
38. 50 X 7 = 350
39. 87+2 = 89
40. 6 X 7 = 42
41. 999 + 89 = 1088
42. 22 X 22 = 484
43. 65 + 92 = 157
44. 1 X 33 = 33
45. 856 X 8 = 6848

Square:

46. $1^2=1$
47. $33^2=1089$
48. $30^2= 900$
49. $17^2=289$
50. $11^2 =121$
51. $13^2 = 169$
52. $51^2=2601$
53. $60^2=3600$
54. $18^2=324$
55. $80^2=6400$

Remainder:

56. 31%6 = 1
57. 185 % 5 = 0
58. 73%8 = 1
59. 73%2=1
60. 31%9 = 4
61. 18%5 = 3

62. 45
63. 177
64. 676
65. 22
66. 70
67. 270

68.

 (42)

69.

 (424)

70.

 (121)

71.

 (4)

72.

 (2023)

Loops

Challenge 73

The answer to the challenge is below. As we can see from the above visual, the value of the counter i is printed every time. The counter is incremented by 1 till it reaches a value of 20.

The loop stops when i reaches its end value of 20.

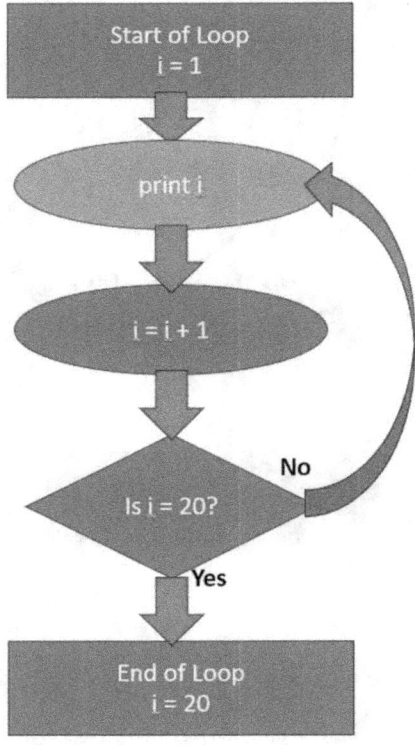

Challenge 74

The answer to the challenge is below. As we can see from the visual, the value of the counter i is printed every time. The counter is incremented by 2 till it exceeds a value of 30.

The loop stops when i is greater than 30.

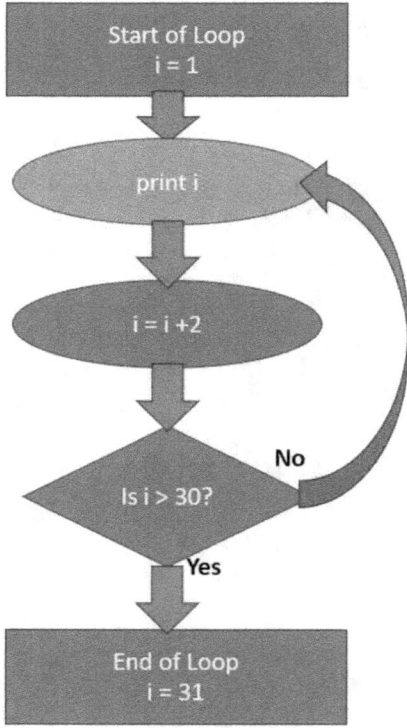

Challenge 75

The answer is below again. The visual below shows that each time we race the car around the track, we increase the counter by 1. The counter in this case is named **carloop**. Once **carloop** reaches 10, the program stops and there's no need to race the car around the track anymore.

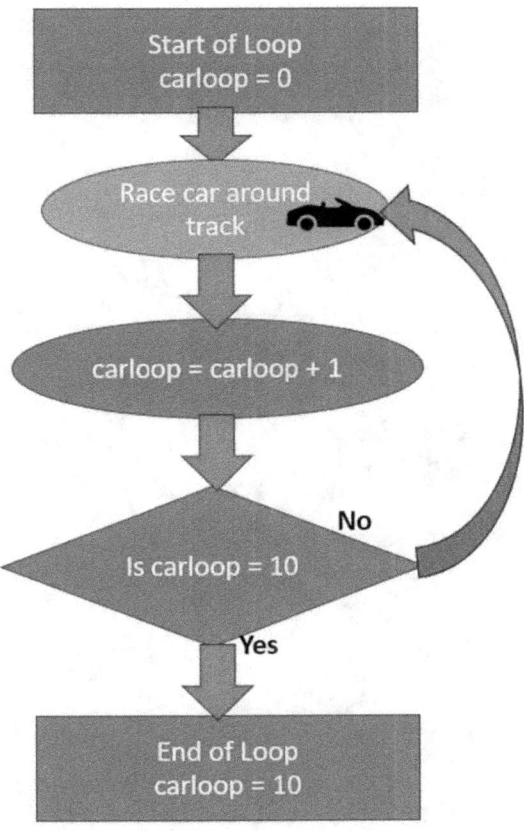

Challenge 76

The answer is below again. The visual below shows that each time the spacecraft orbits Mars, we increase the counter by 1. The counter in this case is named **marsorbit**. Once **marsorbit** reaches 5, the program stops and there's no need to race the car around the track anymore.

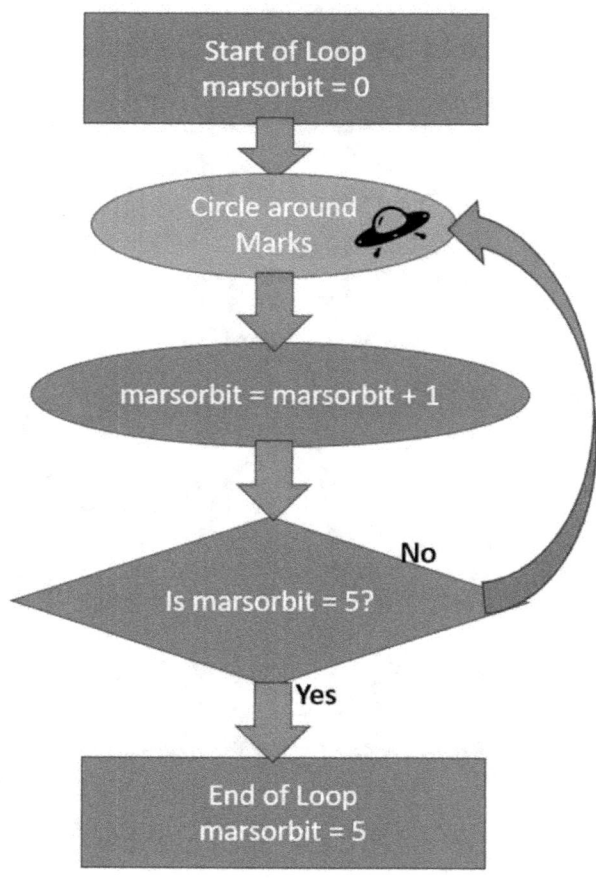

Challenge 77

The answer is below again. The visual below shows that the pattern is printed on the screen each time the counter increases. The counter in this case is named **i**. There is an additional step where the cursor moves to the next line to ensure that the pattern is only printed once each line. Once **i** reaches 100, the program stops and there's no need to race the car around the track anymore.

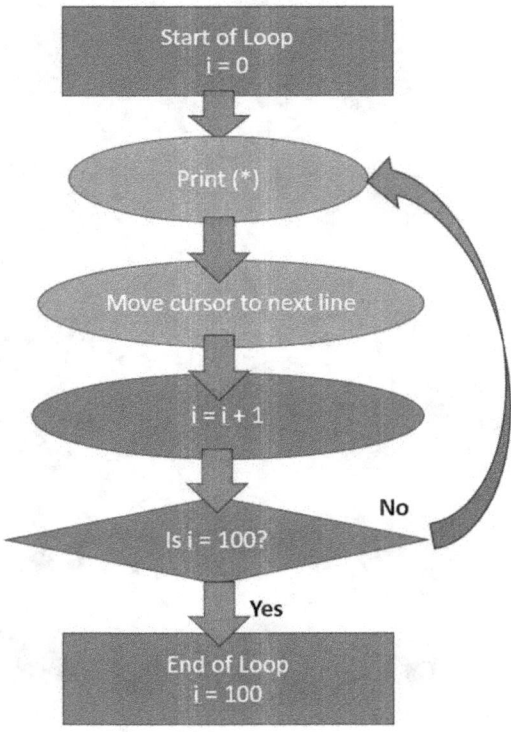

Nested Loops

Challenge 78

The answer to this question is below. The inner loop in green counts the number of bends that happen in each lap. Once the number of bends reaches three, we resume the outer loop and increase the number of laps by 1. This happens till 100 laps are complete.

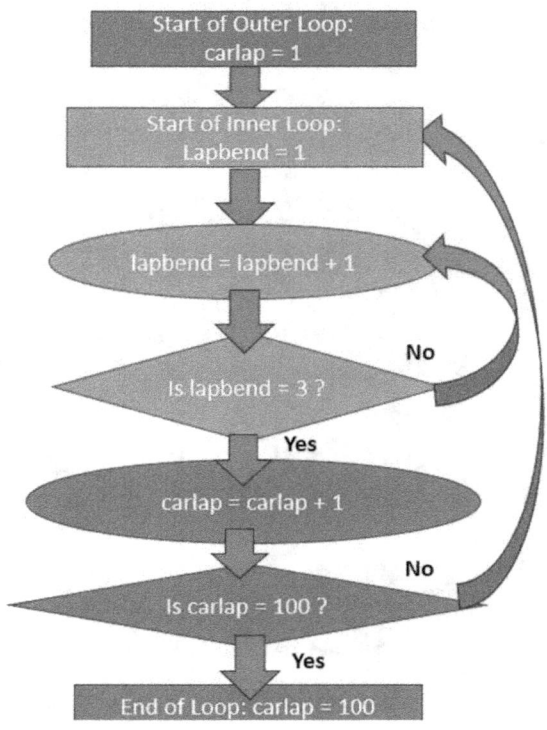

Challenge 79

Output:

1

3

6

10

15

21

28

36

45

55

Challenge 80

The completed flowchart is below. The outer loop has a counter i, which goes from 1 to 10. The counter is incremented by 1 each time and the cursor is moved to the next line.

The green inner loop has a counter j, which goes from 1 to 1. Every time it is incremented by 1, j is printed on the screen to achieve the required pattern.

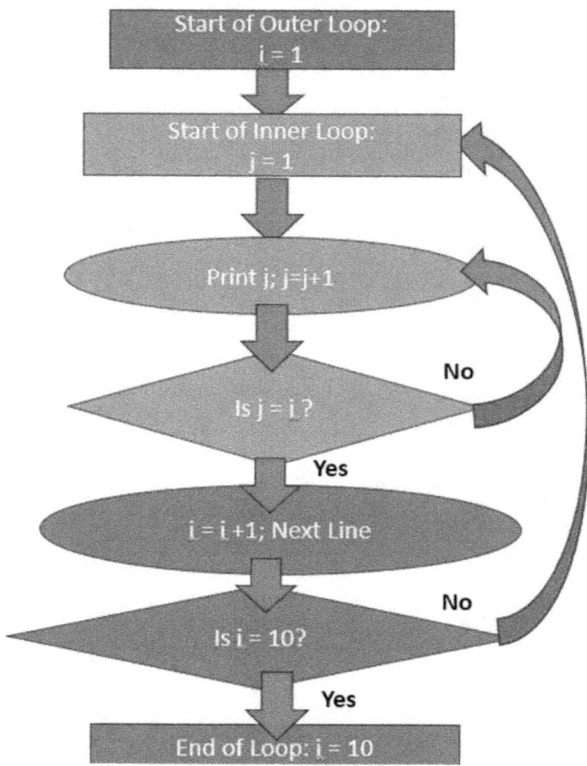

Challenge 81

The answer below is similar to the previous problem. We just print the character * on the screen in the inner loop (instead of the number j).

We do this for 6 iterations of outer loop to achieve 6 lines of output.

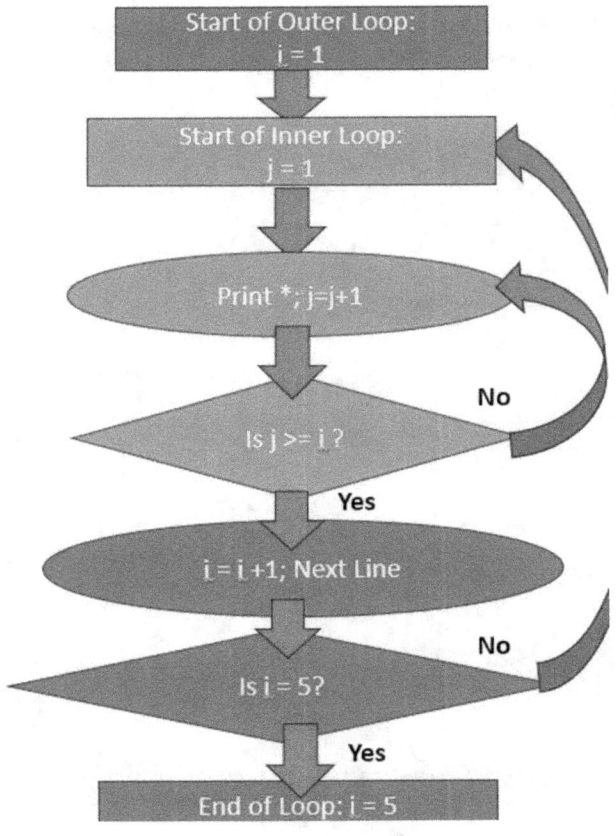

Conditional Loops

Challenge 82

Below is the completed loop. The loop starts with a heads counter of 0. The counter is increased by 1 each time the coin is tossed, and it is heads. Once the heads counter reaches 5, the loop ends. If it is not 5 yet, the loop starts again with another coin toss.

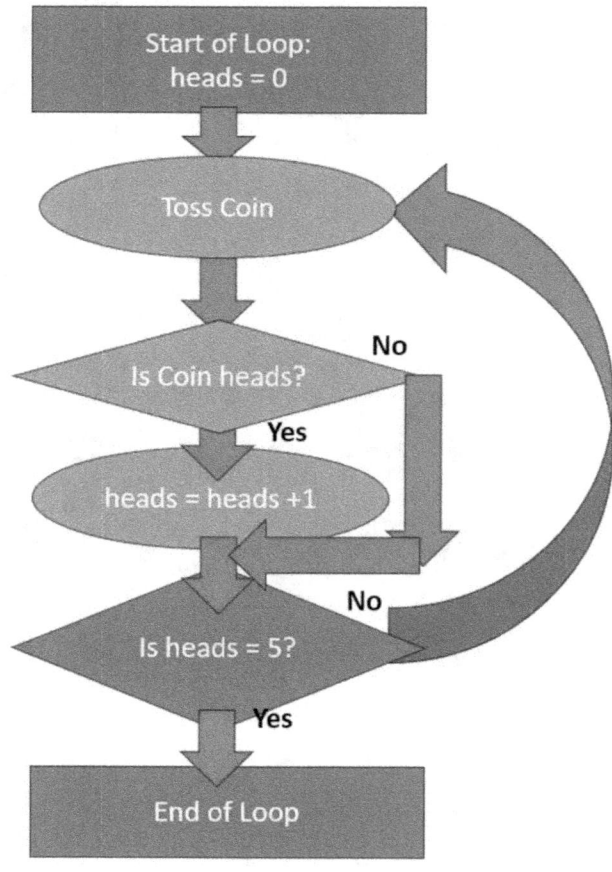

Challenge 83

Below is the completed loop.

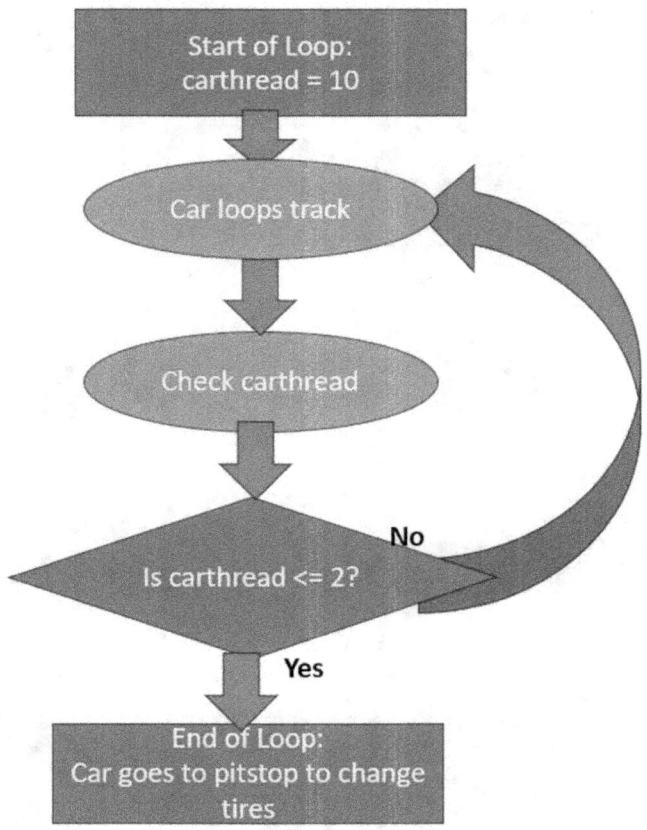

Challenge 84

Below is the completed loop that checks if the credit balance is below $100 after each transaction. If not, the customer is allowed to make additional transactions and then the **transactionamount** is subtracted from **creditbalance**.

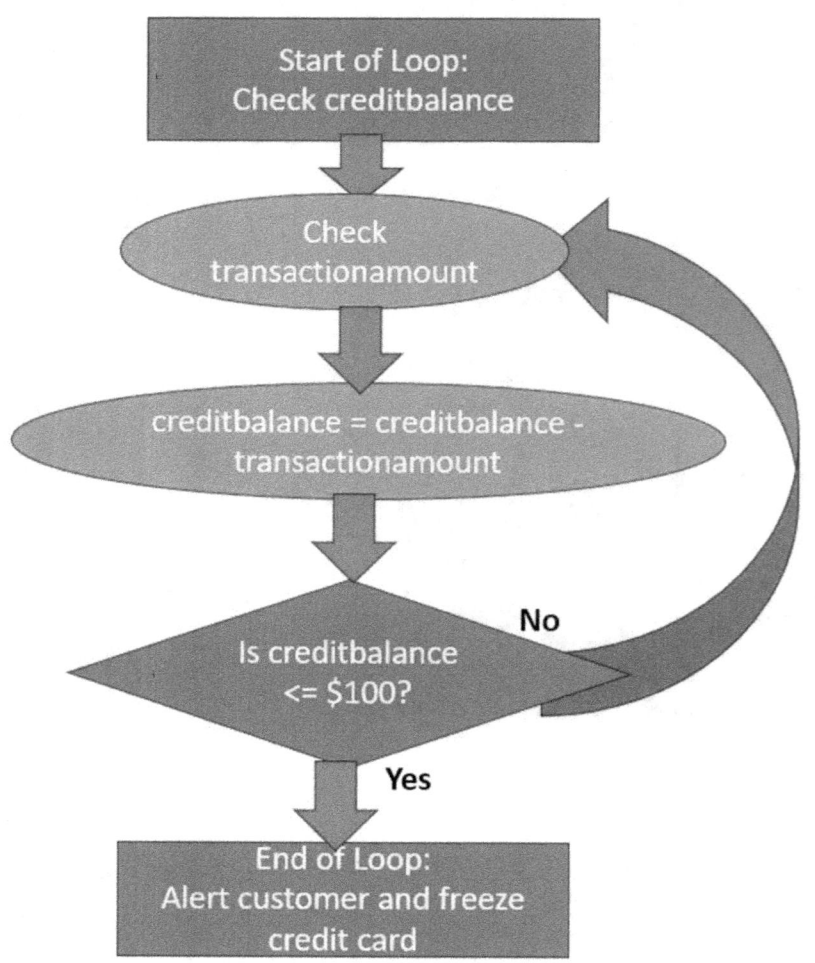

Programming Challenges

Find the outputs for the following loops:

Challenge 85

Output:

/
 /
  /
   /
    /
     /
      /
       /
        /
         /

Challenge 86

Output:

11

22

33

44

55

Challenge 87
Output:

2

4

6

8

10

12

14

16

18

20

Challenge 88

Output:

Sam and Tim are friends.

Challenge 89

Output:

Challenge 90

Output:

Car refuels every 10 laps; ends race after 100 laps.

Challenge 91

Output:

This is a 2D monkey-banana map.

Monkey goes to right if banana is to the right.

Monkey goes left if banana is to the left.

Program ends when monkey reaches banana.

Challenge 92

Output:

This is a 2D monkey-banana map.

Monkey goes up if banana is higher.

Monkey goes down if banana is lower.

Program ends when monkey reaches banana.

Challenge 93

Output:

1

2

6

24

120

720

www.ingramcontent.com/pod-product-compliance
Lightning Source LLC
Chambersburg PA
CBHW071318110526
44591CB00010B/935